Bocconi & Springer Series

Mathematics, Statistics, Finance and Economics

Volume 9

The Bocconi & Springer Series aims to publish research monographs and advanced textbooks covering a wide variety of topics in the fields of mathematics, statistics, finance, economics and financial economics. Concerning textbooks, the focus is to provide an educational core at a typical Master's degree level, publishing books and also offering extra material that can be used by teachers, students and researchers. The series is born in cooperation with Bocconi University Press, the publishing house of the famous academy, the first Italian university to grant a degree in economics, and which today enjoys international recognition in business, economics, and law. The series is managed by an international scientific Editorial Board. Each member of the Board is a top level researcher in his field, well-known at a local and global scale. Some of the Board Editors are also Springer authors and/or Bocconi high level representatives. They all have in common a unique passion for higher, specific education, and for books. Volumes of the series are indexed in Web of Science - Thomson Reuters. Manuscripts should be submitted electronically to Springer's mathematics editorial department: francesca.bonadei@springer.com

THE SERIES IS INDEXED IN SCOPUS

More information about this series at http://www.springer.com/series/8762

Grigorij Kulinich • Svitlana Kushnirenko •
Yuliya Mishura

Asymptotic Analysis
of Unstable Solutions
# of Stochastic Differential
Equations

BOCCONI
UNIVERSITY
PRESS

 Springer

Grigorij Kulinich
Department of General Mathematics
Taras Shevchenko National University
of Kyiv
Kyiv, Ukraine

Svitlana Kushnirenko
Department of General Mathematics
Taras Shevchenko National University
of Kyiv
Kyiv, Ukraine

Yuliya Mishura
Department of Probability Theory, Statistics
and Actuarial Mathematics
Taras Shevchenko National University
of Kyiv
Kyiv, Ukraine

ISSN 2039-1471   ISSN 2039-148X   (electronic)
Bocconi & Springer Series
ISBN 978-3-030-41293-7   ISBN 978-3-030-41291-3   (eBook)
https://doi.org/10.1007/978-3-030-41291-3

This Springer imprint is published by the registered company Springer Nature Switzerland AG.
The registered company address is: Gewerbestrasse 11, 6330 Cham, Switzerland

Preface

This book is devoted to unstable solutions of stochastic differential equations (SDEs). Despite the huge interest in the theory of SDEs, with the help of which the phenomena of nature, technology, economics and finance are modeled, as far as we know this is the first book to present a systematic study of the instability and asymptotic behavior of the corresponding unstable stochastic systems. The book is the result of many years of work by its main co-author, G.L. Kulinich, who devoted a considerable part of his scientific research to the unstable solutions of SDEs. Two other co-authors, S.V. Kushnirenko and Yu.S. Mishura, were very pleased to translate the main ideas of this theory into rigorously stated and clearly proved results, as well as to investigate more general cases of asymptotic behavior of unstable solutions and give relevant examples. The study of the conditions of existence and the asymptotic behavior of unstable solutions, which is proposed herein, started in 1965 when A.V. Skorokhod offered such a problem to his PhD student G.L. Kulinich. At that time the concept of an unstable solution was very poorly investigated; most experts, rather than focusing on instability problems, were attempting to study the conditions of existence and the asymptotics of stable solutions.

This can be explained by the fact that, both in the past and, indeed still today, the problem of the stability of the solution of an SDE, in one or another sense (stability in probability, asymptotic stability in probability, stability in the mean-square sense, exponential stability etc.) is the key problem, important for the various applications. Since stability is an asymptotic property, we are immediately faced with the asymptotic problems which play a leading role in the theory of SDEs, regardless of whether their solutions are stable or not. One of the possible asymptotic approaches is to study stability under random perturbations of the parameters including random perturbations of the noise. In this case, one often considers the situation when the initial condition and noise are asymptotically small, and the stability of the trivial solution is studied, in one or another sense. When we study the asymptotic behavior, as time $t \to +\infty$, of SDE solutions, we are mostly interested in the conditions of "stabilization" of the solution. Among these conditions we have the conditions under which the solution is stable in its direct sense, that is, converges to a certain constant in some stochastic sense, and the conditions under which the solution is ergodic, that is, converges to a certain random variable. This approach is applied to study properties, for example, of mechanical systems subjected to random perturbations and evolution processes involved in finance, economics, medicine, biology, electronics and telecommunications. This and the accompanying problems are discussed in detail in the books [26, 62, 80]; see also references therein. We are not going to review all of the many books and articles on the theory of SDEs, but for an initial acquaintance with this theory and its applications, we suggest that the reader consults the following textbooks and monographs [3, 23, 27, 63, 70, 74].

In contrast to the above books and numerous other works, we are considering the case in which the solution of the SDE is unstable. The notion of instability will be discussed in detail in what follows; however, we prefer to give one definition immediately with the goal of explaining to the reader, albeit briefly, the range of issues under consideration. Note that there can be different definitions of instability, but the following definition is, in our opinion, very simple, clear and easy to check.

Definition 1 A stochastic process $\xi = \{\xi(t), t \geq 0\}$ is called

(i) Stochastically unstable if

$$\lim_{t \to +\infty} \frac{1}{t} \int_0^t \mathbf{P}\{|\xi(s)| < N\}\, ds = 0$$

for any constant $N > 0$.

(ii) Unbounded in probability at infinity if

$$\lim_{t \to +\infty} \mathbf{P}\{|\xi(t)| < N\} = 0$$

for any constant $N > 0$.

For example, the Wiener process is stochastically unstable. Also, it is clear that a stochastically continuous stochastic process is stochastically unstable assuming that it is unbounded in probability. Within our theory, we will apply these concepts to the solutions of Itô's SDEs, which are stochastically continuous processes, even continuous with probability 1. The book itself describes in detail the "unstable" limit theorems and asymptotic properties of solutions. The limit theorems contained in our book are not merely of purely mathematical value; rather, they also have quite practical value. Instability or violations of stability have been noted in many phenomena, and the authors of their descriptions have tried to apply certain mathematical and stochastic methods to deal with them.

Here are a few examples: the unstable states of physical and biological systems, as well as the instability of chemical reactions, have been described in [1, 4, 5, 8, 10, 68, 76, 83]. From our side, let us consider some examples of the behavior of dynamical systems where the instability occurs. For example, a volcanic eruption has occurred, and we are interested in the problem of finding the most likely areas of accumulation of volcanic particles, or in the problem of how to adjust the flow of these particles so that it goes in the necessary direction. If the drift coefficient in the equation has a periodic structure, then the situation arises whereby so-called shaking off takes place and at the same time the structure of the environment is changed. As a very particular example that will, however, be clear to every reader, with intensive shaking of milk, butter and buttermilk are produced; in other words, a bilayer environment is formed.

In general, a bilayer environment naturally appears when we consider a simple stochastic system, e.g., described by a Wiener process, perturbed by some variable intensity $\lambda(x)$. The asymptotic behavior of such system depends on the asymptotics of $\lambda(x)$ at $\pm\infty$ that immediately gives us a bilayer environment. Evidently, ergodic (stable) behavior of the stochastic process does not cover the above anomalies and bundles. Even such a brief description of the applications convinces us that the study of instability is completely logical. Note also that the bilayer environment in our framework is described by so called skew Brownian motion, see, e.g., [2, 21, 57]. It should be mentioned, in particular, that the bilayer environment described by skew Brownian motion, appears in finances. The paper [12] provides evidence of systematic mispricing of the Black–Scholes model when the log-returns of the underlying asset are skewed and leptokurtic. This is described mathematically in [11].

Our main goals achieved in this book are: we have explored Brownian motion in environments with anomalies and we have studied the motion of the Brownian particle in layered media. In the limit we have obtained a fairly wide class of continuous Markov processes. In particular, this class includes Markov processes with discontinuous transition densities. This class includes also processes that are not solutions of any Itô's SDEs, and the Bessel diffusion process.

In addition, in this monograph we study the weak convergence of the additive integral type functionals, under a suitable normalization for unstable solutions of SDEs, to certain functionals of their limit processes.

The book is organized as follows. In Chap. 2 we consider the homogeneous one-dimensional Itô's SDE and obtain necessary and sufficient conditions of weak convergence of the solution to Itô's SDE to Brownian motion in a bilayer environment. For the limit process the explicit form of the transient density is found. In addition, we consider classes of equations for which necessary and sufficient conditions of weak convergence of the normalized solutions to the processes of the skew Brownian motion type are obtained.

We note that from the results of Chap. 2 the principle of spatial averaging of the coefficients of an SDE follows: under certain conditions, in the equation instead of its coefficients we can put constant coefficients which are a certain spatial averaging of the given coefficients. At the same time, the asymptotic form of the distribution for the solution to the SDE does not change.

In Chap. 3 we conduct an analysis of the asymptotic behavior of solutions of equations that are on the verge of equations whose solutions have an ergodic distribution and equations with stochastically unstable solutions. Sufficient conditions for weak convergence of the normalized solutions to the Bessel diffusion process are established for them. The results of Chap. 3 are refinement and generalization of the papers [38, 40, 46].

Chapter 4 focuses on the asymptotic behavior of the additive integral-type functionals under a suitable normalization for unstable solutions of SDEs. These results summarize the papers [34] and [36] and contain the results from [52–54]. New classes of limit distributions, which are certain functionals of the Brownian motion in a bilayer environment or certain functionals of the Bessel diffusion process are obtained. In particular, they contain the local time of the limit processes at the point 0 on the interval $[0, t]$. Similar limit functionals for the Wiener process were first considered in the monograph [81, Chapter 5].

In Chap. 5 we consider homogeneous one-dimensional SDEs with non-regular dependence on parameter. We study the asymptotic behavior of integral functionals of an ordinary Lebesgue integral type, integral functionals of martingale type and mixed functionals defined on the solutions of the equations. We obtain the limit processes in the form of integral functionals of an ordinary Lebesgue integral type, integral functionals of martingale type and mixed functionals with subordinate Wiener processes. These results summarize some results of the paper [40] and contain the results from [55] and [56].

Chapter 6 generalizes some results from the previous chapter to the case of the solutions to inhomogeneous equations with non-regular dependence on the parameter. Under certain conditions, we prove that the asymptotic behavior of the solutions and some functionals of the solutions to inhomogeneous Itô SDEs is the same as that for the solutions to homogeneous Itô SDEs.

All chapters contain a range of examples that illustrate statements about the weak convergence of the solutions and various types of functionals of the solutions to Itô SDEs. To make the book self-contained, definitions and auxiliary results are presented in the Appendix. We include here both the well-known classical theorems concerning the weak convergence of stochastic processes and auxiliary calculations for the main theorems describing the limit behavior of unstable solutions.

The book will be interesting and useful for specialists in stochastic analysis and SDEs, as well as for physicists, chemists, economists, sociologists and other researchers who deal with unstable systems and for practitioners who apply stochastic models to describe phenomena of instability. The basic concepts of this book are quite accessible to graduate students.

We are thankful to everybody who contributed to the improvement of this book. In the earliest stages of his research, G.L. Kulinich was inspired and directed by his teacher A.V. Skorokhod, whose ideas influenced the overall content of the book. Our special thanks are due to Springer Milan and the Editorial Board of the Bocconi & Springer Series for their helpful comments and recommendations which helped to significantly improve the book's presentation.

Kyiv, Ukraine   Grigorij Kulinich
Kyiv, Ukraine   Svitlana Kushnirenko
Kyiv, Ukraine   Yuliya Mishura
2019

Contents

About the Authors

Prof. Grigorij Kulinich received his PhD in probability and statistics from Kyiv University in 1968 and completed his postdoctoral degree in probability and statistics (Habilitation) in 1981. His research work focuses mainly on asymptotic problems of stochastic differential equations with nonregular dependence on parameter, theory of stochastic differential equations, and theory of stochastic processes. He is the author of more than 150 published papers and 3 books.

Dr. Svitlana Kushnirenko is an Associate Professor in the Department of General Mathematics, Taras Shevchenko National University of Kyiv, where she also completed her PhD in probability and statistics in 2006. Her research interests include theory of stochastic differential equations and stochastic analysis. She is the author of 20 papers.

Prof. Yuliya Mishura received her PhD in probability and statistics from Kyiv University in 1978 and completed her postdoctoral degree in probability and statistics (Habilitation) in 1990. She is currently a professor at Taras Shevchenko National University of Kyiv. She is the author/coauthor of more than 270 research papers and 9 books. Her research interests include theory and statistics of stochastic processes, stochastic differential equations, fractional processes, stochastic analysis, and financial mathematics.

Acronyms

Abbreviations

a.s.   Almost surely
a.e.   Almost everywhere
SDE   Stochastic differential equation
w.r.t.   With respect to

Notation

$\mathbb{R}^+$	$[0, +\infty)$
$(\Omega, \mathfrak{F}, \mathbb{F}, \mathsf{P})$	Complete probability space with filtration $\mathbb{F} = \{\mathfrak{F}_t\}_{t \geq 0}$
$\xi = \xi(\omega), \ \omega \in \Omega$	Random variable (a real $\mathfrak{F}$-measurable function)
$\mathsf{E}\xi$	Expectation of a random variable ξ
$W = \{W(t), t \geq 0\}$	Wiener process
χ_A	Indicator function of a set A
$\delta(\cdot)$	Dirac's delta function
$L_\zeta^{x_0}(t)$	Local time of the process ζ at the point x_0 on the interval $[0, t]$
$x \wedge y$	$\min\{x, y\}$
$\xrightarrow{\mathsf{P}}$	Convergence in probability
$\overset{d}{=}$	Equality in distribution
$\langle \zeta \rangle(t)$	Quadratic characteristic of the martingale $\zeta(t)$
$\sigma\{\zeta(s), s \leq t\}$	Smallest σ-algebra such that all random variables $\zeta(s)$ for $s \leq t$ are measurable
$\overset{N}{\underset{-N}{V}} \sigma$	Variation of the function σ on an interval $[-N, N]$

Chapter 1
Introduction to Unstable Processes and Their Asymptotic Behavior

The purpose of this chapter is to introduce the reader to the basic concepts related to unstable processes. To convince the reader that stochastically unstable processes are an important subject for consideration, let us continue with further definitions and visualizations.

Definition 1.1 A solution ξ of Itô's SDE is called ergodic with a distribution function $F(x)$ if

$$\lim_{t \to +\infty} \mathsf{P}\{\xi(t) < x\} = F(x)$$

for any $x \in \mathbb{R}$.

Note that ergodic solutions ξ are not stochastically unstable, since for them there exists the limit in Definition 1, which is not equal to zero for any $N > 0$ (see Lemma A.14 in Appendix).

Let us compare the minor changes in the equations from Figs. 1.1 and 1.2 with the significantly different properties of the solutions of these equations. So, is not it important to be able to make certain changes in stochastically unstable systems (in other words, to control stochastic systems) in order to compensate anomalies and to obtain a system with specified properties?

Now we turn to such (rather interesting and important) cases, when the solution of Itô's SDE is unbounded in probability. In such a case, it is necessary to introduce a suitable normalization, in order to obtain the limit distribution of such a solution. To find non-degenerate limit distributions of unstable solutions ξ to SDEs, we introduce a non-random normalizing factor $B(T) > 0$, $B(T) \to +\infty$, as $T \to +\infty$, where $T > 0$ is a parameter.

© Springer Nature Switzerland AG 2020
G. Kulinich et al., *Asymptotic Analysis of Unstable Solutions of Stochastic Differential Equations*, Bocconi & Springer Series 9,
https://doi.org/10.1007/978-3-030-41291-3_1

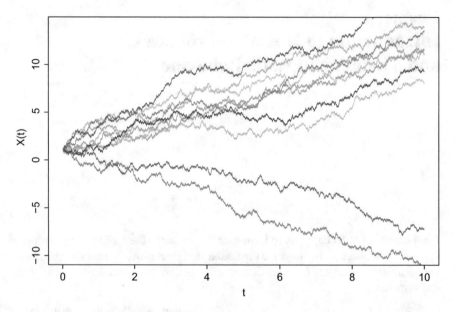

Fig. 1.1 Realizations of stochastically unstable solution to SDE $dX(t) = \text{sign}\, X(t)\, dt + dW(t)$ with initial condition $X(0) = 1$

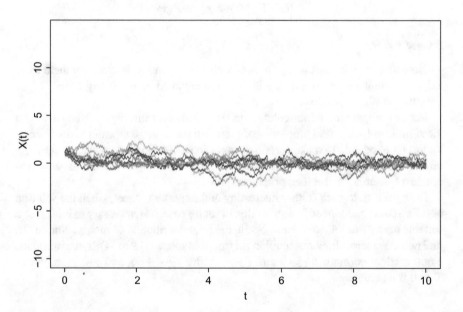

Fig. 1.2 Realizations of ergodic solution to SDE $dX(t) = -\text{sign}\, X(t)\, dt + dW(t)$ with initial condition $X(0) = 1$

Definition 1.2 A solution ξ to Itô's SDE has an exact order of growth $B(t)$, as $t \to +\infty$, if there exists a non-random function $B(t) \to +\infty$, as $t \to +\infty$, and a constant $c_0 \neq 0$ such that

$$\mathsf{P}\left\{\lim_{t \to +\infty} \frac{\xi(t)}{B(t)} = c_0\right\} = 1.$$

It is clear that the solutions to SDEs, which have an exact order of growth $B(t)$, are stochastically unstable. Next we consider the asymptotic behavior of distributions of the processes $\xi_T(t) = \frac{\xi(tT)}{B(T)}$, $t \geq 0$, as $T \to +\infty$.

In addition to the convergence of finite-dimensional distributions of the processes ξ_T, as $T \to +\infty$, we also study the weak convergence to some limit process ζ in the following sense.

Definition 1.3 A family $\xi_T = \{\xi_T(t), t \geq 0\}$ of stochastic processes is said to converge weakly, as $T \to +\infty$, to a process $\zeta = \{\zeta(t), t \geq 0\}$ if, for any $L > 0$, the measures $\mu_T[0, L]$, generated by the processes $\xi_T(\cdot)$ on the interval $[0, L]$ converge weakly to the measure $\mu[0, L]$ generated by the process $\zeta(\cdot)$.

Remark 1.1 Since, as it was mentioned above, the processes ξ_T are continuous with probability 1 as the solutions to Itô's SDEs, Definition 1.3 is a definition of the weak convergence of the processes ξ_T to the continuous process ζ in a uniform topology of the space of continuous functions.

1.1 Equation with the Unit Diffusion Coefficient

1.1.1 Description and Motivation of the Model

Let ξ be the solution of the stochastic differential equation

$$\xi(t) = x_0 + \int_0^t a\left(\xi(s)\right) ds + W(t), \quad t \geq 0, \tag{1.1}$$

where $W = \{W(t), t \geq 0\}$ is a Wiener process, $a = a(x)$ is a continuous, absolutely integrable on the whole axis function with

$$\int_{\mathbb{R}} a(x) dx = \lambda. \tag{1.2}$$

The initial problem from the range of problems considered in this monograph is to study the limiting behavior, as $t \to +\infty$, of the distribution of the process ξ.

To characterize the situation in general, from the point of view of appearance a Wiener process, we can say that it is the case where the deterministic system

$$x(t) = x_0 + \int\limits_0^t a\,(x(s))\,ds$$

is intrinsically perturbed by the stochastic "irregular" process and this perturbation is an important and unavoidable thing. As it is well known, a Wiener process W with probability 1 has no derivative at any point and has unbounded variation on any interval $[t_1,\ t_2]$. So, a deterministic system is considered, whose motion is described by a nonlinear differential equation $\dot{x}(t) = a\,(x(t))$, and which is intrinsically perturbed by a "white noise" process; shortly speaking, the deterministic system is distorted by noise. But, referring to the variety of applications, we can take a different point of view. Namely, the problem (1.1) and (1.2) can be considered as the description of an external perturbation by the coefficient a of the physical environment described by the Wiener process. So, our goal is to study the corresponding state changes of this environment with time. In this case, we solve the problem of non-random control of the structure of a chaotic environment in which a small Brownian particle moves. In particular, under the condition $\lambda \neq 0$ we can assume that at the initial point $x = 0$ an energy source of a high power is implemented in a homogeneous environment.

The main motivation for considering of our model (1.1) and (1.2) is the fact that it can be used in the mathematical description of anomalous phenomena. The book explores Brownian motion in environments with anomalies, that is, environments where certain high power sources can be the energy at certain points. This is, for example, underwater volcanic eruptions or nuclear explosions, tsunamis, tornadoes, and other turbulence, while in the classical case, the motion of a Brownian particle in "smooth" media is considered. Note that the mutual influence of tsunamis, tornadoes, hurricanes, technological disasters with climate has acquired a decisive significance nowadays. Our approach also covers more peaceful phenomena such as the motion of the Brownian particle in layered media, such as, for example, oil and water. Some other examples are provided in Preface.

1.1.2 Asymptotic Growth and Normalizing Multipliers for the Solutions

When studying Eq. (1.1) under the condition (1.2), it turned out that the solution ξ is unbounded in probability, as $t \to +\infty$, the details are contained in [30] for $\lambda = 0$ and in [29] for $\lambda \neq 0$. Therefore, for establishing a non-degenerate boundary distribution of the solution ξ, it was necessary to introduce some non-random normalizing multiplier $B(t) \to +\infty$, as $t \to +\infty$. In this connection, it was necessary to develop new methods for investigating the behavior when $T \to +\infty$ of the distribution of the normalized random process $\xi_T(t) = \frac{\xi(tT)}{B(T)}, t > 0$, where T is parameter. Considering Eq. (1.1) under the condition (1.2), it was noticed that one can chose $B(T) = \sqrt{T}$, and the normalized solution $\xi_T(t)$ satisfies the equation

$$\xi_T(t) = \frac{x_0}{\sqrt{T}} + \int_0^t a_T(\xi_T(s)) \, ds + W_T(t), \quad t \geq 0, \tag{1.3}$$

where $a_T(x) = \sqrt{T}a(x\sqrt{T})$, $W_T(t) = \frac{W(tT)}{\sqrt{T}}$ is a family of Wiener processes. By the way, note that Eq. (1.3) is an integral form of the stochastic differential Itô equation

$$d\xi_T(t) = a_T(\xi_T(t)) \, dt + dW_T(t), \quad t \geq 0. \tag{1.4}$$

Let us also emphasize that the coefficient $a_T(x)$ of Eq. (1.3) contains the multiplier $\sqrt{T}$ that tends to infinity, as $T \to +\infty$, therefore the standard asymptotic methods are not suitable for the investigation of the limit behavior of the solutions $\xi_T(t)$ of Eq. (1.3), as $T \to +\infty$. In particular, in the case where $\lambda \neq 0$, $a_T(x)$ is a "δ"-shaped family of functions, whose "δ"-shaped property is realized at the point $x = 0$ and it has the weight $\lambda = \int_{\mathbb{R}} a(x) \, dx$. In this connection, the cases where $\lambda = 0$ and $\lambda \neq 0$ in (1.2) are essentially different concerning the limit behavior of the distributions of the solutions $\xi_T(t)$ to Eq. (1.3). To specify the above description of the situation with butter and buttermilk, we can say that in the case where $\lambda \neq 0$, and a has a periodic structure, our Eq. (1.3) contains the coefficient $a_T(x) = \sqrt{T}a(x\sqrt{T})$ that corresponds to "intensive shaking" and to creation of a bilayer limit environment. Concerning the case $\lambda = 0$, a new asymptotic method of the investigation of the "inconvenient" term $\int_0^t a_T(\xi_T(s)) \, ds$ was proposed in the paper [30]. This method applies the representation

$$\int_0^t a_T(\xi_T(s)) \, ds = \Phi_T(\xi_T(t)) - \Phi_T(x_0) - \int_0^t \Phi'_T(\xi_T(s)) \, dW_T(s), \tag{1.5}$$

where

$$\Phi_T(x) = 2 \int_0^x \left[1 - \sigma_0 e^{-2\int_0^u a_T(v) \, dv} \right] du, \quad \sigma_0 = e^{2\int_0^{+\infty} a(z) \, dz}.$$

As one can see, the right-hand side of the representation (1.5) contains a stochastic Itô integral. According to the Itô formula, the representation (1.5) holds with probability 1 for any $t > 0$. Moreover, to obtain (1.5), we use the equality

$$\Phi_T'(x)a_T(x) + \frac{1}{2}\Phi_T''(x) = a_T(x),$$

that holds for any $x \in \mathbb{R}$. Therefore, the investigation of the limit behavior of the "inconvenient" term $\int_0^t a_T(\xi_T(s))\,ds$ can be reduced to the investigation of two things: first, the limit behavior of the functional consisting of the function $\Phi_T(x)$, where the value $\xi_T(t)$ is substituted, and, second, the asymptotic behavior of the family of martingales

$$\eta_T(t) = \int_0^t \Phi_T'(\xi_T(s))\,dW_T(s).$$

Recall that the functions $\Phi_T(x)$ and $\Phi_T'(x)$ contain the coefficient $a_T(x)$ of Eq. (1.3) under the sign of an integral. This fact simplifies essentially the asymptotic study of the behavior of the solutions $\xi_T(t)$. When studying the asymptotic problems connected to Eq. (1.1) under the condition (1.2) with $\lambda = 0$, this method allows to get the convergence

$$\xi_T(t) - W_T(t) \to 0,$$

as $T \to +\infty$, in probability, for any $t > 0$. Therefore, in the case where the space-average value of the function a is zero, i.e., $\int_{\mathbb{R}} a(x)\,dx = 0$, we have that

$$\frac{1}{\sqrt{t}}\int_0^t a(\xi(s))\,ds \to 0$$

in probability, as $t \to +\infty$. In this case $a = a(x)$ does not affect the limit behavior of the distribution of the solution ξ of Eq. (1.1), as $t \to +\infty$.

Later on, this result was obtained in the monograph [17], also via the analysis of the right-hand side of the representation (1.5), however, the convergence of the right-hand side to zero was established by different methods.

The case where $\lambda \neq 0$ in (1.1) and (1.2) differs essentially from the case $\lambda = 0$. According to [29], it turned out that in this case the value

$$\frac{1}{\sqrt{t}}\int_0^t a(\xi(s))\,ds \nrightarrow 0$$

in probability, as $t \to +\infty$. In view of this fact, another new probabilistic method was proposed. This method is to deal with the transformation $\zeta_T(t) = f_T(\xi_T(t))$ of the solution $\xi_T(t)$ of Eq. (1.3). Such a transformation "annihilates" the term $\int_0^t a_T(\xi_T(s))\, ds$ when choosing f_T properly. It means that it is reasonable to consider harmonic functions $f_T(x)$ that solve the equation

$$f_T'(x) a_T(x) + \frac{1}{2} f_T''(x) = 0$$

for any $x \in \mathbb{R}$. In particular, we can take the functions $f_T(x)$ as follows:

$$f_T(x) = \int_0^x \exp\left\{-2\int_0^u a_T(v)\, dv\right\} du, \quad x \in \mathbb{R} \tag{1.6}$$

and investigate the behavior of the distributions of the process $\zeta_T(t) = f_T(\xi_T(t))$, where $\xi_T(t)$ is the solution of Eq. (1.3). According to the Itô formula,

$$\zeta_T(t) = f_T(x_0) + \int_0^t \widehat{\sigma}_T(\zeta_T(s))\, dW_T(s), \tag{1.7}$$

where $\widehat{\sigma}_T(x) = f_T'(\varphi_T(x))$, and $\varphi_T(x)$ are the inverse functions to the functions $f_T(x)$ for any $T > 0$. According to condition (1.2), there exist constants $\delta > 0$ and $C > 0$ such that $0 < \delta \le f_T'(x) \le C$ for any $x \in \mathbb{R}$. Therefore, the investigation of the behavior of the distributions of the processes $\zeta_T(t)$ is reduced to the investigation of the behavior of the martingales $\int_0^t \widehat{\sigma}_T(\zeta_T(s))\, dW_T(s)$. Further, we can apply A.V. Skorokhod subsequence convergence theorem (see Theorem A.12 or [79, Chapter I, § 6]) which states that for some subsequence $T_n \to +\infty$ we have the convergence in probability of the subsequences $\zeta_{T_n}(t) \to \zeta(t)$ and $W_{T_n}(t) \to W(t)$ for any $t > 0$.

1.1.3   Bilayer Environment and Transition Density of the Limit Homogeneous Markov Process

Condition (1.2) provides an opportunity to establish the relationship

$$\int_0^t P\{\zeta(s) = 0\}\, ds = 0 \tag{1.8}$$

for any $t > 0$, and this in turn allows one to come to the limit in (1.7) and get that the limit process $\zeta(t)$ satisfies the equation

$$\zeta(t) = \int_0^t \overline{\sigma}(\zeta(s))dW(s), \qquad (1.9)$$

where $\overline{\sigma}(x) = \begin{cases} \sigma_1, & x \geq 0, \\ \sigma_2, & x < 0, \end{cases}$ $\sigma_1 = \exp\{-2 \int_0^{+\infty} a(x)\,dx\}$, $\sigma_2 = \exp\{-2 \int_0^{-\infty} a(x)\,dx\}$.

By virtue of equality (1.8) the integral in (1.9) is an Itô integral, which means that Eq. (1.9) is an Itô stochastic differential equation and ζ is a weak solution of this equation. Since for any $T > 0$ the process ζ_T is a homogeneous Markov process, see [79], the limit process ζ is a homogeneous Markov process as well. Applying the equality (1.8), the explicit form of the transition density $\rho(t, x, y)$ of the process ζ was established in the paper [29]. This transition density has a discontinuity of the first kind, i.e., a jump, at the point $y = 0$, and this phenomena appears as the result of adhesion at zero of two Gaussian distributions: $\mathbb{N}(0, \sigma_1^2 t)$, restricted to the positive semi-axis, and $\mathbb{N}(0, \sigma_2^2 t)$, restricted to the negative semi-axis, correspondingly. Therefore, the process ζ describes the projection on the real axis of the trajectory of a Brownian particle, moving in a bilayer environment with the boundary at the point $y = 0$, subject to the continuous passage of the boundary by this particle. In this case, a certain refraction of the trajectory occurs, similar to the refraction of the trajectory of the light beam as it passes through the boundary of two media. Moreover, due to the convergence $f_T(x) \to x\overline{\sigma}(x)$ and $\varphi_T(x) \to \frac{x}{\overline{\sigma}(x)}$ for any $x \neq 0$, as $T \to +\infty$, and the inequality $0 < \delta_1 \leq \varphi'_T(x) \leq C_1$ that holds for any x and T with some constants δ_1 and C_1 not depending on x and T, we get the weak convergence, as $T \to +\infty$, of the solution ξ_T of Eq. (1.3) to the homogeneous Markov process $\widehat{\xi}(t) = l(\zeta(t))$. Here $\zeta(t)$ is the solution of Eq. (1.9), $l(x) = \frac{x}{\overline{\sigma}(x)}$. Therefore, the transition density of the process $\widehat{\xi}(t)$ has the form

$$\rho_{\widehat{\xi}}(t, x, y) = \rho_\zeta\left(t, l^{-1}(x), l^{-1}(y)\right)\left(l^{-1}(y)\right)', \qquad (1.10)$$

where $l^{-1}(x) = x\overline{\sigma}(x)$ is the inverse function to the function $l(x)$.

Let us emphasize that the process $\widehat{\xi}$ describes quite adequately the trajectory of motion of a Brownian particle in a homogeneous environment having a barrier at the point $y = 0$. In this case the symmetry which is inherent in the Wiener process is broken. More precisely, the symmetry of probability of hitting the set $y > 0$ and the set $y < 0$, starting from the point $y = 0$, is broken. For a Wiener process such probabilities equal $\frac{1}{2}$, while for the process $\widehat{\xi}$ they equal $\frac{\sigma_2}{\sigma_2 + \sigma_1}$ and $\frac{\sigma_1}{\sigma_2 + \sigma_1}$, respectively (see [17, Chapter 3, § 15, Theorem 4]). Therefore, (1.1) and (1.2) with $\lambda \neq 0$ can be applied to construct a mathematical model of an atomic explosion leading to multidirectional streams of radioactive particles. Note that in the book [24] the process $\widehat{\xi}$ was called a skew Brownian motion. Besides this, it was established by M.I. Portenko in the book [72] that the diffusion process $\widehat{\xi}$ admits Kolmogorov local characteristics in the generalized sense. This fact served him as the basis for introducing the notion of a generalized diffusion process that is a homogeneous Markov process, admitting Kolmogorov local characteristics in the generalized sense. For example, the process $\widehat{\xi}$ admits a generalized drift coefficient of the form $c\delta(\cdot)$, where $\delta(\cdot)$ is Dirac's delta function, $c = \frac{\sigma_2 - \sigma_1}{\sigma_2 + \sigma_1} = \tanh\lambda$, and the diffusion coefficient equals 1.

Therefore, if we consider the problem (1.1) under the condition (1.2) with $\lambda \neq 0$, then the process ξ_T converges weakly to the process $\widehat{\xi}$, which is a generalized process in the sense described above, and for which one can formally write the stochastic differential equation

$$d\widehat{\xi}(t) = \tanh\lambda\,\delta(\cdot)\,dt + dW(t), \tag{1.11}$$

where $\delta(\cdot)$ is Dirac's delta function at the point zero.

Compare this situation to the case when $\lambda = 0$. In this case the solutions ξ_T of Eq. (1.3) converge weakly to a Wiener process, i.e., $\widehat{\xi} = W$.

Note that A. Friedman [15] established that in the problem (1.1) and (1.2) the condition $\lambda = 0$ is necessary for the convergence, as $T \to +\infty$, of the distributions of the process $\xi_T(t) = \frac{\xi(tT)}{\sqrt{T}}$ to a Wiener process W. Besides this, it follows from [42] that for the process $\widehat{\xi}$ the equality

$$\widehat{\xi}(t) = \beta(t) + W(t)$$

holds, where $\beta(t)$ is some additive functional depending on the solution ζ of Eq. (1.9).

1.1.4 Comparison with the Smooth Disturbing Process

We stress again that in this way, the situation with the limit behavior, as $T \to +\infty$, of the finite-dimensional distributions of the process $\xi_T(t) = \frac{\xi(tT)}{\sqrt{T}}$, where ξ is the solution of the problem (1.1) and (1.2), depends on the value of λ. In other words, it depends on the limit behavior of the coefficient $a(x)$, as $|x| \to +\infty$. This fact is connected to the non-regular behavior of the disturbing process W. If the disturbing process is smooth, the situation is different. For example, consider the stationary process in the wide sense $\eta = \{\eta(t), t \geq 0\}$, where $\mathsf{E}\eta(t) = 0$ and the random variables $\eta(t_1)$ and $\eta(t_2)$ are independent for $|t_2 - t_1| \geq l > 0$ and any t_i. It was proved in [32] that finite-dimensional distributions of the process $\xi_T(t) = \frac{\xi(tT)}{\sqrt{T}}$, where ξ is the solution of the equation

$$\xi(t) = x_0 + \int_0^t a\,(\xi(s))\,ds + \int_0^t \eta(s)\,ds, \quad t \geq 0,$$

and condition (1.2) holds, converge, as $T \to +\infty$, to the finite-dimensional distributions of the process $\sigma_0 W$, where W is a Wiener process, $\sigma_0 = \int_{-l}^{l} R(u)\,du$ and $R(u)$ is the covariance function of the process η. It means that the limit finite-dimensional distributions of the process $\xi_T(t)$, as $T \to +\infty$, do not depend on λ.

1.1.5   Spatial Averaging of the Vibrational Type Coefficient

Undoubtedly, the question on the influence of the coefficient a of a vibrational type, e.g., $a(x) = \sin x$, on the behavior of the solution ξ_T, as $T \to +\infty$, is very intriguing. For the first time such a problem for Eq. (1.1) was studied in [31] using the methods developed in the papers [29] and [30]. In particular, the results of this paper demonstrate that the existence of the spatial averaging of a of the form

$$
\frac{1}{x}\int_{0}^{x} \exp\left\{-2\int_{0}^{u} a(v)\,dv\right\} du \to \sigma_1, \quad
\frac{1}{x}\int_{0}^{x} \exp\left\{2\int_{0}^{u} a(v)\,dv\right\} du \to \sigma_2,
$$

(1.12)

as $|x| \to +\infty$, $0 < \sigma_i < +\infty$, supplies that the finite-dimensional distributions of the process $\xi_T(t) = \frac{\xi(tT)}{\sqrt{T}}$, $t > 0$, where ξ is the solution of (1.1), converge, as $T \to +\infty$, to the finite-dimensional distributions of the process $\sigma_0 W$, where W is a Wiener process, $\sigma_0 = (\sigma_1 \sigma_2)^{-\frac{1}{2}}$. In the case where $a(x) = \sin x$, we have that $\sigma_0 = \left[\sum_{n=0}^{+\infty} \frac{1}{(n!)^2}\right]^{-1}$, or, in other words, $\frac{1}{\sigma_0}$ is the so-called Bessel's constant. It is clear that here we have $\sigma_0 < \frac{1}{2}$, and the coefficient a "eats" in the limit some part of the coefficient of the Wiener process W, or, in other words, coefficient a changes in the limit the structure of the environment, the mathematical model of which is described by Eq. (1.1).

We note that the principle of spatial averaging of the coefficients of the stochastic differential equations was first substantiated in the paper [31]. This principle means that the existence of the spatial averaging (1.12) in Eq. (1.1) allows to substitute the averaged constants, instead of the coefficients of the equation, and get the equation

$$
\eta(t) = \frac{1}{\sqrt{\sigma_1 \sigma_2}} W(t).
$$

Herewith the limit distributions of the solutions coincide, i.e., $\frac{\xi(t)}{\sqrt{t}} \sim \frac{\eta(t)}{\sqrt{t}}$, as $t \to +\infty$.

Let us mention that the problem (1.1) and (1.2) with $\lambda \neq 0$ was studied in [75], using purely analytical methods. More precisely, the limiting behavior of the characteristic operators of the Markov processes was investigated.

1.1.6 Functionals of the Solution

We ask the reader to pay attention to the fact that by investigating the problem (1.1) and (1.2), we deal with the study of the limiting behavior at $+\infty$ of the distribution of the functional $\frac{1}{\sqrt{t}} \int_0^t a\left(\xi(s)\right) ds$ depending on the solution ξ of Eq. (1.1). The significance of the investigation of the limit behavior, at $+\infty$, of integral functionals of the solutions of SDE is very well known (see, e.g., [81]). The limit behavior of the functionals of integral type of unstable solutions ξ of Eq. (1.1) was first reviewed in [34] and [36], whereby a method was used which is based on the representation of the type (1.5).

Furthermore, the problem of the asymptotic behavior of the finite-dimensional distributions of the solution of Eq. (1.3) is very important in the case where the external disturbance is non-regular, but is a Wiener process only in the limit. For example, consider the equations of the form

$$\xi_T(t) = x_0 + \int_0^t a_T\left(\xi_T(s)\right) ds + \eta_T(t), \tag{1.13}$$

where $\eta_T(t)$ is a family of a.s. continuous, square-integrable martingales with quadratic characteristic $\langle \eta_T \rangle(t) \overset{P}{\to} t$, as $T \to +\infty$. It means that a Wiener process in Eq. (1.13) appears only in the limit.

The asymptotic behavior of the finite-dimensional distributions of the solutions ξ_T of Eq. (1.13) and of the respective integral functionals, in the case where $\eta_T(t) \neq W_T(t)$, was first obtained in the paper [39]. In this case it is necessary to impose additional conditions that connect with the rate of degeneracy of the coefficients $a_T = a_T(x)$ at some points, for example, if $|a_T(x_k)| \to +\infty$, as $T \to +\infty$, and with the rate of convergence $\langle \eta_T \rangle(t) - t \to 0$, under the same condition $T \to +\infty$.

Not less important is the discrete analogue of the limit theorems for unstable solutions. The first results in this direction were obtained in the paper [37].

1.2 Equation with the Non-unit Diffusion Coefficient

For a deeper understanding of the essence of the established facts concerning the influence of the coefficients on the limit behavior of the distribution of the solution of Eq. (1.1), it is reasonable to consider Eq. (1.1) as a mathematical model of the physical phenomenon of diffusion, or as the model of the trajectory of the motion of a small Brownian particle, the mass of which we can neglect and which is in a liquid or in gas and moves under the influence of micro- and macroscopic factors of the environment. However, the coefficient which determines its motion, so-called diffusion coefficient, can differ from being simply 1. In such a more general case, when the motion is guided by some nontrivial diffusion coefficient, it is known that the projection on the axis of the trajectory of the motion of the small Brownian particle in the environment, homogeneous in time, is well described by the SDE

$$\xi(t) = x_0 + \int_0^t a\left(\xi(s)\right) ds + \int_0^t \sigma\left(\xi(s)\right) dW(s), \tag{1.14}$$

where $\int_0^t a\left(\xi(s)\right) ds$ is an ordinary Lebesgue integral, $\int_0^t \sigma\left(\xi(s)\right) dW(s)$ is a stochastic Itô integral, and W, as usual, is a Wiener process defined on some probability space $(\Omega, \mathfrak{F}, \mathsf{P})$, $\mathfrak{F}_t = \sigma\{W(s), s \leq t\}$. Recall that W describes the diffusion in the homogeneous environment. This means that a more general class of stochastic differential equations appears on the scene. Obviously, (1.14) with $\sigma(x) \equiv 1$ coincides with Eq. (1.1). Equation (1.14) can be written in the differential Itô form

$$d\xi(t) = a\left(\xi(t)\right) dt + \sigma\left(\xi(t)\right) dW(t), \quad \xi(0) = x_0. \tag{1.15}$$

Discretizing, we can say that the drift $\Delta\xi(t) = \xi(t + \Delta t) - \xi(t)$ of the Brownian particle on a comparatively small time interval Δt has the form

$$\Delta\xi(t) \simeq a\left(\xi(t)\right) \Delta t + \sigma\left(\xi(t)\right) \Delta W(t),$$

where $\Delta W(t) = W(t + \Delta t) - W(t)$ is the drift of the Brownian particle in a motionless homogeneous environment. So, $a(x)$ is a macroscopic characteristic of the environment, in other words, the velocity of its movement calculated at the point x, while $\sigma^2(x)$ is a microscopic characteristic of the environment, in other words, the intensity of thermal chaotic motion of the environment's molecules at the point x. Therefore coefficient a in Eq. (1.15) is called drift coefficient, while σ is called diffusion coefficient of the environment. Standard conditions

(I$_1$)   there exists a constant $L > 0$ such that for any $x \in \mathbb{R}$

$$|a(x)|^2 + |\sigma(x)|^2 \leq L(1 + |x|^2);$$

(I$_2$)   for any $N > 0$ there exist constants $L_N > 0$, $\delta_N > 0$ such that $0 < \delta_N \leq \sigma(x) \leq L_N$ for $|x| \leq N$

imply that Eq. (1.15) has a weak solution (ξ, W), and this solution is weakly unique. Moreover, ξ is a homogeneous process having the strong Markov property [80].

Under the conditions (I$_1$), (I$_2$) and

(I$_3$)   for any $N > 0$ there exists a constant $L_N > 0$ such that the variation $\overset{N}{\underset{-N}{V}} \sigma(x) \leq L_N$, and $\sigma(x) \geq \sigma_0 > 0$ for any $x \in \mathbb{R}$,

Eq. (1.15) has a strong solution ξ, and this solution is strongly unique [18]. Various situations concerning existence and uniqueness of the strong and weak solutions are described in the Introduction to the book [9]. Concerning the present book, sufficient conditions for the weak convergence, as $T \to +\infty$, were obtained for integral functionals of unstable solutions ξ of Eq. (1.15) and real-valued function $g = g(x)$. Namely, they are established for

$$\beta_T^{(1)}(t) = \frac{1}{B_1(T)} \int_0^{tT} g\left(\xi(s)\right) ds, \qquad \beta_T^{(2)}(t) = \frac{1}{B_2(T)} \int_0^{tT} g\left(\xi(s)\right) dW(s), \quad t \geq 0,$$

where $B_i(T)$ are normalizing constants, and $B_i(T) \to +\infty$, as $T \to +\infty$, (ξ, W) is a weak solution of Eq. (1.15). New classes of limit distributions are obtained that are distributions of the following functionals

$$\beta^{(1)}(t) = 2 \left[\int_0^{\zeta(t)} \bar{b}(u)\, du - \int_0^{t} \bar{b}(\zeta(s))\, d\zeta(s) \right], \quad \beta^{(2)}(t) = \xi\sqrt{\beta^{(1)}(t)},$$

where, $\bar{b}(x) = b_1$ for $x \geq 0$ and $\bar{b}(x) = b_2$ for $x < 0$, $\zeta(t)$ is the strong solution of the Itô equation

$$d\zeta(t) = \bar{\sigma}(\zeta(t))\, dW(t).$$

Here $\bar{\sigma}(x) = \sigma_1$ for $x \geq 0$ and $\bar{\sigma}(x) = \sigma_2$ for $x < 0, 0 < \sigma_i, i = 1, 2, \xi$ is a standard Gaussian random variable, independent of $\zeta(t)$. Previously, such limit distributions were obtained in the book [81] for $\zeta(t) = \widehat{W}(t)$, where $\widehat{W}$ is a Wiener process, in the process of studying additive functionals of random walk.

Now we mention, for the convenience of the readers, some related results, which either complement the theorems proved in the book, or can serve as illustrations and applications. Namely, in the paper [35], a consistent non-asymptotically normal drift parameter estimator for the Itô SDE is constructed. The law of the iterated logarithm is established for the one-dimensional diffusion process in [43]. The asymptotic behavior of the solution of the Cauchy problem for parabolic equations of the second order with non-regular dependence of the coefficients on a parameter is investigated in [49] in a nonclassical case, where the pointwise convergence of the coefficients of the equation is not required. An analysis of the asymptotic behavior of a harmonic oscillator with external perturbation by random processes of the "white and shot noises" types is carried out in [44]. As a kind of the pioneer result, an analog of the well-known arcsine law for a Wiener process is obtained in [7] for a Brownian motion in a bilayer environment.

In the paper [46], the asymptotic behavior of the one-dimensional Itô SDE with non-regular dependence of the coefficients on a parameter is investigated. In this case, the convergence (in a certain sense) of the drift coefficient to functions of the form $c_k|x - x_k|^{-\alpha}$, where $\alpha > 0$, is assumed. In [66], a method for studying the convergence rate of the normalized unstable SDEs' solutions to the limit process is proposed. In [51], the results of the paper [37] are generalized in the sense that the weak convergence of a sequence of Markov chains to a diffusion type process is obtained.

Similar problems for unstable solutions of systems of SDEs are considered in the paper [15]. In [50], an exact order of growth and the law of the iterated logarithm for a part of components of the solutions for SDEs in $\mathbb{R}^n$ are obtained. In the paper [38], the weak convergence for a part of normalized unstable components of the solutions for SDEs in $\mathbb{R}^n$ to the solutions of some SDE with discontinuous coefficients or to the Bessel diffusion process is considered. The paper [47] investigates the asymptotic behavior of distributions for normalized unstable solutions of some systems of SDEs with Poisson jumps. The weak convergence for the solutions of SDEs to the discontinuous processes in the Skorokhod topology is considered in [45]. The asymptotic behavior of functionals of the solutions to inhomogeneous Itô stochastic differential equations with non-regular dependence on a parameter is studied in [48].

For modern issues of existence, uniqueness of solutions, their properties and asymptotic analysis, all this in the context of singularity and instability, we refer the readers to the papers [6, 13, 58, 58–64, 71]. Instability in the dynamical system involving fractional Brownian motion is considered in [16].

Chapter 2
Convergence of Unstable Solutions of SDEs to Homogeneous Markov Processes with Discontinuous Transition Density

In this chapter we consider one-dimensional homogeneous stochastic differential equations with stochastically unstable solutions. Conditions on the coefficients of the equations leading to instability of the solutions are established in Sect. 2.1. Necessary and sufficient conditions for the weak convergence of the stochastically unstable solutions to a Brownian motion in two-layer environment are formulated and proved in Sect. 2.2. Necessary and sufficient conditions for the weak convergence of stochastically unstable solutions of SDEs to a process of skew Brownian motion type are obtained in Sect. 2.3. Section 2.4 contains several examples that illustrate statements about the weak convergence of the stochastically unstable solutions. Auxiliary results are collected in Appendix A.

2.1 Preliminaries

Let the real-valued measurable functions $a = a(x)$ and $\sigma = \sigma(x) : \mathbb{R} \to \mathbb{R}$ be given. Assume that a and σ satisfy the following conditions:

(i) a and σ are of linear growth, i.e., the following inequality holds: for some $L > 0$ and any $x \in \mathbb{R}$

$$|a(x)| + |\sigma(x)| \leq L(1 + |x|);$$

(ii) for any $N > 0$ there exists $\delta_N > 0$ such that $\sigma(x) \geq \delta_N > 0$ for $|x| \leq N$.

Also, let $(\Omega, \mathfrak{F}, \mathbb{F}, \mathsf{P})$ be a complete probability space with filtration $\mathbb{F} = \{\mathfrak{F}_t\}_{t \geq 0}$ satisfying the standard assumptions, and let $W = \{W(t), t \geq 0\}$ be a one-dimensional Wiener process which is adapted to the filtration $\mathbb{F}$.

© Springer Nature Switzerland AG 2020
G. Kulinich et al., *Asymptotic Analysis of Unstable Solutions of Stochastic Differential Equations*, Bocconi & Springer Series 9,
https://doi.org/10.1007/978-3-030-41291-3_2

Consider the stochastic differential equation (SDE)

$$d\xi(t) = a\left(\xi(t)\right) dt + \sigma\left(\xi(t)\right) dW(t), \quad t \geq 0 \tag{2.1}$$

with $\mathfrak{F}_0$-measurable initial condition $\xi(0)$.

The function a is called the drift coefficient, and the function σ is called the diffusion coefficient of the SDE (2.1).

Note that everywhere in the book where we refer to Eq. (2.1), we assume that the conditions (i)–(ii) are satisfied.

Definition 2.1 A strong solution of Eq. (2.1) is a progressively measurable process $\{\xi(t), t \geq 0\}$ satisfying the integral equation

$$\xi(t) = \xi(0) + \int_0^t a\left(\xi(s)\right) ds + \int_0^t \sigma\left(\xi(s)\right) dW(s)$$

a.s. for any $t \geq 0$.

Here $\int_0^t a\left(\xi(s)\right) ds$ is a Lebesgue integral, and $\int_0^t \sigma\left(\xi(s)\right) dW(s)$ is a stochastic Itô integral.

Definition 2.2 A weak solution to Eq. (2.1) is a triple consisting of the following components:

– a stochastic basis $(\Omega', \mathfrak{F}', \{\mathfrak{F}'_t\}_{t \geq 0}, \mathsf{P}')$;
– a Wiener process W' on this basis;
– an adapted process $\xi' = \{\xi'(t), t \geq 0\}$ on this basis such that $\xi'(0) \overset{d}{=} \xi(0)$ and

$$d\xi'(t) = a\left(\xi'(t)\right) dt + \sigma\left(\xi'(t)\right) dW'(t), \quad t \geq 0.$$

If for any two solutions ξ_1 and ξ_2 of Eq. (2.1), their finite-dimensional distributions coincide, then the solution ξ of Eq. (2.1) is called weakly unique. It means that we have uniqueness of the solution in distribution. If with probability 1 the paths for any two solutions ξ_1 and ξ_2 of Eq. (2.1) coincide, then the solution of Eq. (2.1) is called strongly unique, that is, we have a pathwise uniqueness. It is known [80, Chapter 1, § 2.1] that a weakly unique solution ξ of Eq. (2.1) is a homogeneous strong Markov process.

Remark 2.1 The existence of a weak solution of Eq. (2.1) was first proved in [79], Chapter 3, §3 under the conditions of continuity and linear growth of the coefficients. Later in the paper [28] the existence of a weak solution and its weak uniqueness were proved under the condition $\sigma(x) \geq \delta > 0$ and the condition of boundedness of coefficients. The existence of a weak solution and its weak uniqueness follow from the paper [28] under conditions (i)–(ii).

An overview of the basic definitions and facts related to different types of existence and uniqueness of the solution to SDEs is contained in the monograph [9], see also [65].

Remark 2.2 On the one hand, it is clear that a strong solution of Eq. (2.1) is of course also a weak solution, and the strong uniqueness of the solution implies weak uniqueness. On the other hand, to study the asymptotic behavior, as $t \to +\infty$, of the distribution of the solution, it is sufficient to assume the existence of a weak solution ξ to Eq. (2.1). In this connection, we will not distinguish between weak and strong solutions in the problems concerning the distribution of solutions.

It should also be emphasized that the asymptotic behavior, as $t \to +\infty$, of solutions to SDE (2.1) is closely related to the behavior, as $|x| \to +\infty$, of the function

$$f(x) = \int\limits_0^x \exp\left\{-2\int\limits_0^u \frac{a(v)}{\sigma^2(v)} \, dv\right\} du. \tag{2.2}$$

Evidently, f' is a positive function satisfying the equation

$$f'(x)a(x) + \frac{1}{2}f''(x)\sigma^2(x) = 0 \tag{2.3}$$

a.e. with respect to the Lebesgue measure. The role of this function has been studied in detail in [17, § 16].

Since we consider stochastically unstable solutions, our goal is to formulate the conditions for the stochastic instability of the solutions to SDEs.

Lemma 2.1 *Let ξ be a solution to Eq. (2.1) and let, additionally to (i)–(ii), the following assumption hold for some constant $C > 0$ and for all $x \in \mathbb{R}$:*

$$0 < f'(x)\,\sigma(x) \le C. \tag{2.4}$$

Then for any constant $N > 0$

$$\lim_{t \to +\infty} \frac{1}{t} \int\limits_0^t \mathbf{P}\{|\xi(s)| < N\} \, ds = 0,$$

i.e., the solution of Eq. (2.1) is stochastically unstable.

Proof Consider the function

$$\Phi_N(x) = 2\int\limits_0^x f'(u)\left(\int\limits_0^u \frac{\chi_{|v|\le N}}{f'(v)\sigma^2(v)} \, dv\right) du,$$

where χ_A is the indicator of the set A. Function $\Phi_N(x)$ has the continuous first derivative

$$\Phi'_N(x) = 2f'(x) \int_0^x \frac{\chi_{|v|\leq N}}{f'(v)\sigma^2(v)}\, dv.$$

Also, its second derivative

$$\Phi''_N(x) = 2f''(x) \int_0^x \frac{\chi_{|v|\leq N}}{f'(v)\sigma^2(v)}\, dv + 2f'(x)\frac{\chi_{|x|\leq N}}{f'(x)\sigma^2(x)}$$

exists a.e. with respect to the Lebesgue measure and is locally integrable. Taking into account the relation (2.3), we obtain

$$\Phi'_N(x)\, a(x) + \frac{1}{2}\Phi''_N(x)\, \sigma^2(x) = \chi_{|x|\leq N} \tag{2.5}$$

a.e. with respect to the Lebesgue measure.

Applying the Itô formula from Lemma A.3 to the process $\Phi_N(\xi(t))$, where ξ is a solution to Eq. (2.1), and equality (2.5), we conclude that

$$\Phi_N(\xi(t)) - \Phi_N(\xi(0)) = \int_0^t \left[\Phi'_N(\xi(s))\, a(\xi(s)) + \frac{1}{2}\Phi''_N(\xi(s))\, \sigma^2(\xi(s))\right] ds$$

$$+ \int_0^t \Phi'_N(\xi(s))\, \sigma(\xi(s))\, dW(s).$$

Therefore, we have

$$\int_0^t \chi_{|\xi(s)|\leq N}\, ds = \Phi_N(\xi(t)) - \Phi_N(\xi(0)) - \int_0^t \Phi'_N(\xi(s))\, \sigma(\xi(s))\, dW(s).$$

Recall that according to (2.4) we have $0 < f'(x)\,\sigma(x) \leq C$. Also, according to linear growth of a (assumption (i)) and to assumption (ii), both functions $f'(v)$ and $\sigma^2(v)$ are separated from zero on the interval $[-N, N]$ by some positive constant that depends on N. As a result, both functions $\int_0^u \frac{\chi_{|v|\leq N}}{f'(v)\sigma^2(v)}\, dv$ and $\Phi'_N(x)\,\sigma(x)$ are bounded, and so

$$\mathsf{E}\left(\int_0^t \Phi'_N(\xi(s))\, \sigma(\xi(s))\, dW(s)\right) = 0.$$

Furthermore,

$$\Phi_N(x) - \Phi_N(y) = 2 \int\limits_y^x f'(u) \left(\int\limits_0^u \frac{\chi_{|v| \leq N}}{f'(v)\sigma^2(v)} \, dv \right) du,$$

and taking into account the boundedness of the positive integral $\int\limits_0^u \frac{\chi_{|v| \leq N}}{f'(v)\sigma^2(v)} dv$ by some constant depending on N, we can deduce that

$$|\Phi_N(x) - \Phi_N(y)| \leq C_N \left| \int\limits_y^x f'(u) du \right| = C_N |f(x) - f(y)|.$$

So, we can bound from above the desirable probability as follows:

$$\frac{1}{t} \int\limits_0^t P\{|\xi(s)| < N\} \, ds = \frac{1}{t} E\left[\Phi_N\left(\xi(t)\right) - \Phi_N\left(\xi(0)\right)\right]$$

$$\leq C_N \frac{1}{t} E |f\left(\xi(t)\right) - f\left(\xi(0)\right)| \leq C_N \left(\frac{1}{t^2} E |f\left(\xi(t)\right) - f\left(\xi(0)\right)|^2 \right)^{\frac{1}{2}}.$$

Applying the Itô formula to the process $f(\xi(t))$, we conclude that

$$f\left(\xi(t)\right) - f\left(\xi(0)\right) = \int\limits_0^t \left[f'(\xi(s)) \, a(\xi(s)) + \frac{1}{2} f''(\xi(s)) \, \sigma^2(\xi(s)) \right] ds$$

$$+ \int\limits_0^t f'(\xi(s)) \, \sigma(\xi(s)) \, dW(s) = \int\limits_0^t f'(\xi(s)) \, \sigma(\xi(s)) \, dW(s).$$

Taking into account the property of stochastic integrals and the assumptions of the lemma, we have

$$E |f\left(\xi(t)\right) - f\left(\xi(0)\right)|^2 = \int\limits_0^t E\left[f'\left(\xi(s)\right) \sigma\left(\xi(s)\right) \right]^2 ds \leq Ct.$$

Therefore, we conclude that

$$\lim_{t \to +\infty} \frac{1}{t} \int\limits_0^t P\{|\xi(s)| < N\} \, ds \leq \lim_{t \to +\infty} C_N \sqrt{\frac{C}{t}} = 0,$$

and the solution ξ to Eq. (2.1) is stochastically unstable. $\qquad \square$

2.2   Necessary and Sufficient Conditions for the Weak Convergence of Solutions of SDEs to a Brownian Motion in a Bilayer Environment

Let $T > 0$ be a parameter. Consider the processes

$$\zeta_T(t) = \frac{f(\xi(tT))}{\sqrt{T}}, \quad W_T(t) = \frac{W(tT)}{\sqrt{T}}, \tag{2.6}$$

where the function $f(x)$ is defined by (2.2), and the processes $\xi(t)$ and $W(t)$ are related via Eq. (2.1). Evidently, for any fixed $T > 0$ the process $W_T = \{W_T(t), t \geq 0\}$ is a Wiener process.

Let $\varphi(x)$ be the inverse function to $f(x)$, and

$$\widehat{\sigma}_T(x) = f'\left(\varphi\left(x\sqrt{T}\right)\right)\sigma\left(\varphi\left(x\sqrt{T}\right)\right), \quad Q_T(x) = \int_0^x \widehat{\sigma}_T^{-2}(u)\, du. \tag{2.7}$$

Note that the function Q_T exists under the conditions (i)–(ii).

Also, let ζ be a solution of the SDE

$$\zeta(t) = \int_0^t \overline{\sigma}(\zeta(s))\, dW(s), \tag{2.8}$$

where

$$\overline{\sigma}(x) = \begin{cases} \sigma_1, & x \geq 0, \\ \sigma_2, & x < 0, \end{cases} \quad 0 < \sigma_i < +\infty.$$

Now, let $c_1 > 0$ and $c_2 > 0$ be two constants. Denote

$$\widehat{\xi}(t) = l(\zeta(t)), \tag{2.9}$$

where

$$l(x) = \frac{x}{\overline{c}(x)}, \quad \overline{c}(x) = \begin{cases} c_1, & x \geq 0, \\ c_2, & x < 0. \end{cases}$$

Remark 2.3 The process (2.9) with $\overline{c}(x) = \overline{\sigma}(x)$ is called the skew Brownian motion (see [24, Section 4.2, Problem 1]). Therefore the process (2.9) will be called the process of the skew Brownian motion type. See Appendix A.3 for more details.

Remark 2.4 The solution ζ to Eq. (2.8) is a homogeneous strong Markov process with the transition density

$$
\rho(t, x, y) = \begin{cases}
\dfrac{1}{\sigma_1\sqrt{2\pi t}}\left[e^{-\frac{(y-x)^2}{2\sigma_1^2 t}} - \dfrac{\sigma_1-\sigma_2}{\sigma_1+\sigma_2} e^{-\frac{(y+x)^2}{2\sigma_1^2 t}} \right], & x \geq 0,\, y > 0, \\[4mm]
\dfrac{2\sigma_1}{\sigma_1+\sigma_2} \cdot \dfrac{1}{\sigma_2\sqrt{2\pi t}} e^{-\frac{(\sigma_1 y - \sigma_2 x)^2}{2\sigma_1^2\sigma_2^2 t}}, & x \geq 0,\, y < 0, \\[4mm]
\dfrac{2\sigma_2}{\sigma_1+\sigma_2} \cdot \dfrac{1}{\sigma_1\sqrt{2\pi t}} e^{-\frac{(\sigma_2 y - \sigma_1 x)^2}{2\sigma_1^2\sigma_2^2 t}}, & x \leq 0,\, y > 0, \\[4mm]
\dfrac{1}{\sigma_2\sqrt{2\pi t}}\left[e^{-\frac{(y-x)^2}{2\sigma_2^2 t}} - \dfrac{\sigma_2-\sigma_1}{\sigma_1+\sigma_2} e^{-\frac{(y+x)^2}{2\sigma_2^2 t}} \right], & x \leq 0,\, y < 0.
\end{cases}
\tag{2.10}
$$

See Appendix A.3 for a proof of this fact. Note that the transition density of the solution ζ to Eq. (2.8) was first obtained in [29].

In what follows constants C and L do not depend on x and T. Now, we are in a position to prove the following statement.

Lemma 2.2 *Let ξ be a solution to Eq. (2.1) and let, additionally to (i)–(ii) and (2.4), the following assumption hold for some constant $C > 0$ and for all $x \in \mathbb{R}$:*

$$
\frac{1}{f(x)} \int_0^x \frac{du}{f'(u)\sigma^2(u)} \leq C.
\tag{2.11}
$$

Then there exist some constant $\widehat{C} > 0$ such that for any $\varepsilon > 0$ and any $t \geq 0$ the following inequality holds:

$$
\int_0^t P\{|\zeta_T(s)| < \varepsilon\}\, ds \leq \varepsilon\, \widehat{C}\sqrt{t}.
\tag{2.12}
$$

Proof Indeed, applying the Itô formula to the process $\Phi_1(\zeta_T(t))$, where

$$
\Phi_1(x) = 2 \int_0^x \left(\int_0^u \widehat{\sigma}_T^{-2}(v)\chi_{(-\varepsilon,\varepsilon)}(v)\, dv \right) du,
$$

we obtain similarly to Lemma 2.1 that

$$
\int_0^t P\{|\zeta_T(s)| < \varepsilon\}\, ds = E\left[\Phi_1(\zeta_T(t)) - \Phi_1(\zeta_T(0))\right]
$$

$$
\leq 2\int_{-\varepsilon}^{\varepsilon} \frac{dv}{\widehat{\sigma}_T^2(v)} E\,|\zeta_T(t) - \zeta_T(0)|
$$

$$= 2\varepsilon \left[\frac{1}{f(\varphi(\varepsilon\sqrt{T}))} \int\limits_{0}^{\varphi(\varepsilon\sqrt{T})} \frac{1}{f'(u)\sigma^2(u)} \, du \right.$$

$$\left. + \frac{1}{f(\varphi(-\varepsilon\sqrt{T}))} \int\limits_{0}^{\varphi(-\varepsilon\sqrt{T})} \frac{1}{f'(u)\sigma^2(u)} \, du \right] \mathsf{E} \, |\zeta_T(t) - \zeta_T(0)|$$

$$\leq 4\varepsilon C \mathsf{E} \, |\zeta_T(t) - \zeta_T(0)| \, .$$

From the last equality, using the boundedness of the function $\widehat{\sigma}_T(x)$ and the inequality

$$\mathsf{E} \, |\zeta_T(t) - \zeta_T(0)|^2 = \mathsf{E} \left(\int\limits_{0}^{t} \widehat{\sigma}_T(\zeta_T(s)) \, dW_T(s) \right)^2 = \int\limits_{0}^{t} \mathsf{E}\widehat{\sigma}_T^2(\zeta_T(s)) \, ds \leq C^2 t,$$

we have (2.12) for $\widehat{C} = 4C^2$.   □

Theorem 2.1 *Let ξ be a solution of Eq. (2.1) and let, additionally to (i)–(ii), there exist a constant $C > 0$ such that, for all $x \in \mathbb{R}$, the inequalities (2.4) and (2.11) hold. Then the stochastic process $\zeta_T = \{\zeta_T(t), t \geq 0\}$ from (2.6) converges weakly, as $T \to +\infty$, to the process ζ satisfying (2.8) if and only if*

$$\frac{1}{f(x)} \int\limits_{0}^{x} \frac{du}{f'(u)\sigma^2(u)} \to \begin{cases} \frac{1}{\sigma_1^2}, & x \to +\infty, \\ \frac{1}{\sigma_2^2}, & x \to -\infty. \end{cases} \tag{2.13}$$

Proof *Sufficiency* Let convergence (2.13) hold. Note that the functions $f(x)$ and $f'(x)$ are continuous, the function $f''(x)$ is locally integrable and we have equality (2.3) a.e. with respect to the Lebesgue measure. Therefore, we can apply the Itô formula from Lemma A.3 to the process $\xi(tT)$ and obtain

$$f(\xi(tT)) - f(\xi(0)) = \int\limits_{0}^{tT} \left[f'(\xi(s)) \, a(\xi(s)) + \frac{1}{2} f''(\xi(s)) \, \sigma^2(\xi(s)) \right] ds$$

$$+ \int\limits_{0}^{tT} f'(\xi(s)) \, \sigma(\xi(s)) \, dW(s) = \int\limits_{0}^{tT} f'(\xi(s)) \, \sigma(\xi(s)) \, dW(s).$$

Therefore, we have

$$\zeta_T(t) = \zeta_T(0) + \frac{1}{\sqrt{T}} \int\limits_{0}^{t} f'(\xi(sT)) \, \sigma(\xi(sT)) \, dW(sT).$$

Taking into account the equality $\xi(tT) = \varphi(f(\xi(tT))) = \varphi\left(\zeta_T(t)\sqrt{T}\right)$, we proceed as follows:

$$\zeta_T(t) = \zeta_T(0) + \int_0^t \widehat{\sigma}_T(\zeta_T(s))\,dW_T(s). \tag{2.14}$$

Since the functions $f'(x)\sigma(x)$ and, consequently, $\widehat{\sigma}_T(x)$ are bounded, we have the following relations:

$$\lim_{C\to+\infty}\ \lim_{T\to+\infty}\ \sup_{0\le t\le L}\ \mathsf{P}\{|\zeta_T(t)| > C\} = 0,$$

$$\lim_{h\to 0}\ \limsup_{T\to+\infty}\ \sup_{|t_1-t_2|\le h,\ t_i<L}\ \mathsf{P}\{|\zeta_T(t_1) - \zeta_T(t_2)| > \varepsilon\} = 0$$

for any $L > 0$ and $\varepsilon > 0$. It is clear that such relations hold for the process W_T as well.

According to Theorem A.12, given an arbitrary sequence $T'_n \to +\infty$, we can choose a subsequence $T_n \to +\infty$, a probability space $(\widetilde{\Omega}, \widetilde{\mathfrak{F}}, \widetilde{\mathsf{P}})$, and a sequence $\left(\widetilde{\zeta}_{T_n}(t),\ \widetilde{W}_{T_n}(t)\right)$ of stochastic processes, defined on this space and such that their finite-dimensional distributions coincide with those of the processes $\left(\zeta_{T_n}(t),\ W_{T_n}(t)\right)$ and, moreover, $\widetilde{\zeta}_{T_n}(t) \xrightarrow{\widetilde{\mathsf{P}}} \widetilde{\zeta}(t)$, $\widetilde{W}_{T_n}(t) \xrightarrow{\widetilde{\mathsf{P}}} \widetilde{W}(t)$, as $T_n \to +\infty$, for all $t \ge 0$, where $\widetilde{\zeta}(t)$, $\widetilde{W}(t)$ are some stochastic processes.

It follows from the weak equivalence of the processes $\left(\zeta_{T_n}(t),\ W_{T_n}(t)\right)$ and the processes $\left(\widetilde{\zeta}_{T_n}(t),\ \widetilde{W}_{T_n}(t)\right)$, combined with Eq. (2.14), Lemma A.11 (see Corollary A.2) and Lemma A.13, that

$$\widetilde{\zeta}_{T_n}(t) = \widetilde{\zeta}_{T_n}(0) + \int_0^t \widehat{\sigma}_{T_n}\left(\widetilde{\zeta}_{T_n}(s)\right)\,d\widetilde{W}_{T_n}(s). \tag{2.15}$$

Now we are in a position to establish that

$$\int_0^t \widehat{\sigma}_{T_n}^2\left(\widetilde{\zeta}_{T_n}(s)\right)ds - \int_0^t \overline{\sigma}^2\left(\widetilde{\zeta}(s)\right)ds \xrightarrow{\widetilde{\mathsf{P}}} 0, \tag{2.16}$$

as $T_n \to +\infty$, where the function $\overline{\sigma}(x)$ is defined by (2.8). Consider the function

$$\Phi_{T_n}(x) = 2\int_0^x \left[u - \int_0^u \frac{\overline{\sigma}^2(v)}{\widehat{\sigma}_{T_n}^2(v)}\,dv\right]du.$$

It satisfies the following relations:

$$\Phi'_{T_n}(x) = 2\left[x - \int\limits_0^x \frac{\overline{\sigma}^2(v)}{\widehat{\sigma}^2_{T_n}(v)}\, dv\right], \quad \Phi''_{T_n}(x) = 2\frac{\widehat{\sigma}^2_{T_n}(x) - \overline{\sigma}^2(x)}{\widehat{\sigma}^2_{T_n}(x)}$$

a.e. with respect to the Lebesgue measure.

The functions $\Phi_{T_n}(x)$ and $\Phi'_{T_n}(x)$ are continuous in x for every T_n, function $\Phi''_{T_n}(x)$ is locally integrable. Therefore, we can apply the Itô formula from Lemma A.3 to the process $\Phi_{T_n}\left(\widetilde{\zeta}_{T_n}(t)\right)$ and obtain

$$\Phi_{T_n}\left(\widetilde{\zeta}_{T_n}(t)\right) - \Phi_{T_n}\left(\widetilde{\zeta}_{T_n}(0)\right) = \frac{1}{2}\int\limits_0^t \Phi''_{T_n}\left(\widetilde{\zeta}_{T_n}(s)\right)\widehat{\sigma}^2_{T_n}\left(\widetilde{\zeta}_{T_n}(s)\right)\, ds$$

$$+ \int\limits_0^t \Phi'_{T_n}\left(\widetilde{\zeta}_{T_n}(s)\right)\widehat{\sigma}_{T_n}\left(\widetilde{\zeta}_{T_n}(s)\right)\, d\widetilde{W}_{T_n}(s)$$

$$= \int\limits_0^t \left[\widehat{\sigma}^2_{T_n}\left(\widetilde{\zeta}_{T_n}(s)\right) - \overline{\sigma}^2\left(\widetilde{\zeta}_{T_n}(s)\right)\right] ds + \int\limits_0^t \Phi'_{T_n}\left(\widetilde{\zeta}_{T_n}(s)\right)\widehat{\sigma}_{T_n}\left(\widetilde{\zeta}_{T_n}(s)\right)\, d\widetilde{W}_{T_n}(s).$$

Therefore, we have

$$\int\limits_0^t \widehat{\sigma}^2_{T_n}\left(\widetilde{\zeta}_{T_n}(s)\right)\, ds - \int\limits_0^t \overline{\sigma}^2\left(\widetilde{\zeta}_{T_n}(s)\right)\, ds$$

$$= \Phi_{T_n}\left(\widetilde{\zeta}_{T_n}(t)\right) - \Phi_{T_n}\left(\widetilde{\zeta}_{T_n}(0)\right) - \int\limits_0^t \Phi'_{T_n}\left(\widetilde{\zeta}_{T_n}(s)\right)\widehat{\sigma}_{T_n}\left(\widetilde{\zeta}_{T_n}(s)\right)\, d\widetilde{W}_{T_n}(s).$$

$$(2.17)$$

Since

$$\left|\Phi'_{T_n}(x)\right| = 2\left|x - \overline{\sigma}^2(x)\frac{x}{f\left(\varphi\left(x\sqrt{T_n}\right)\right)}\int\limits_0^{\varphi(x\sqrt{T_n})} \frac{dv}{f'(v)\,\sigma^2(v)}\right|,$$

according to the condition (2.13), we have for some constant $C > 0$ and for any constant $N > 0$

$$\left|\Phi'_{T_n}(x)\right| \le C\,|x|, \qquad \lim_{T_n \to +\infty}\ \sup_{|x|\le N}\left|\Phi'_{T_n}(x)\right| = 0.$$

Consider the set

$$B = \left\{ \sup_{0 \le s \le t} \left| \widetilde{\zeta}_{T_n}(s) \right| \le N \right\} \quad \text{and let} \quad P_N = \mathsf{P} \left\{ \sup_{0 \le s \le t} \left| \widetilde{\zeta}_{T_n}(s) \right| > N \right\} = \mathsf{P} \left\{ B^c \right\}.$$

For any $\varepsilon > 0$ and $N > 0$ we have the following inequalities:

$$\mathsf{P} \left\{ \left| \Phi_{T_n} \left(\widetilde{\zeta}_{T_n}(t) \right) - \Phi_{T_n} \left(\widetilde{\zeta}_{T_n}(0) \right) \right| > \varepsilon \right\}$$

$$= \mathsf{P} \left\{ \left| \Phi_{T_n} \left(\widetilde{\zeta}_{T_n}(t) \right) - \Phi_{T_n} \left(\widetilde{\zeta}_{T_n}(0) \right) \right| (\chi_B + \chi_{B^c}) > \varepsilon \right\}$$

$$\le \mathsf{P} \left\{ \left| \Phi_{T_n} \left(\widetilde{\zeta}_{T_n}(t) \right) - \Phi_{T_n} \left(\widetilde{\zeta}_{T_n}(0) \right) \right| \chi_B > \frac{\varepsilon}{2} \right\}$$

$$+ \mathsf{P} \left\{ \left| \Phi_{T_n} \left(\widetilde{\zeta}_{T_n}(t) \right) - \Phi_{T_n} \left(\widetilde{\zeta}_{T_n}(0) \right) \right| \chi_{B^c} > \frac{\varepsilon}{2} \right\}$$

$$\le \mathsf{P} \left\{ \left| \Phi_{T_n} \left(\widetilde{\zeta}_{T_n}(t) \right) - \Phi_{T_n} \left(\widetilde{\zeta}_{T_n}(0) \right) \right| \chi_B > \frac{\varepsilon}{2} \right\} + \mathsf{P} \left\{ \sup_{0 \le s \le t} \left| \widetilde{\zeta}_{T_n}(s) \right| > N \right\}$$

$$\le P_N + \frac{2}{\varepsilon} \mathsf{E} \left(\left| \Phi_{T_n} \left(\widetilde{\zeta}_{T_n}(t) \right) - \Phi_{T_n} \left(\widetilde{\zeta}_{T_n}(0) \right) \right| \chi_B \right)$$

$$\le P_N + \frac{4N}{\varepsilon} \sup_{|x| \le N} \left| \Phi'_{T_n}(x) \right|,$$

$$\mathsf{P} \left\{ \left| \int_0^t \Phi'_{T_n} \left(\widetilde{\zeta}_{T_n}(s) \right) \widehat{\sigma}_{T_n} \left(\widetilde{\zeta}_{T_n}(s) \right) d\widetilde{W}_{T_n}(s) \right| > \varepsilon \right\}$$

$$\le \mathsf{P} \left\{ \left| \int_0^t \Phi'_{T_n} \left(\widetilde{\zeta}_{T_n}(s) \right) \widehat{\sigma}_{T_n} \left(\widetilde{\zeta}_{T_n}(s) \right) \chi_B \, d\widetilde{W}_{T_n}(s) \right| > \frac{\varepsilon}{2} \right\}$$

$$+ \mathsf{P} \left\{ \left| \int_0^t \Phi'_{T_n} \left(\widetilde{\zeta}_{T_n}(s) \right) \widehat{\sigma}_{T_n} \left(\widetilde{\zeta}_{T_n}(s) \right) \chi_{B^c} \, d\widetilde{W}_{T_n}(s) \right| > \frac{\varepsilon}{2} \right\} = I_1 + I_2.$$

We apply Chebyshev's inequality and obtain

$$I_1 \le \left(\frac{2}{\varepsilon} \right)^2 \mathsf{E} \left[\int_0^t \Phi'_{T_n} \left(\widetilde{\zeta}_{T_n}(s) \right) \widehat{\sigma}_{T_n} \left(\widetilde{\zeta}_{T_n}(s) \right) \chi_B \, d\widetilde{W}_{T_n}(s) \right]^2$$

$$= \frac{4}{\varepsilon^2} \int_0^t \mathsf{E} \left[\Phi'_{T_n} \left(\widetilde{\zeta}_{T_n}(s) \right) \widehat{\sigma}_{T_n} \left(\widetilde{\zeta}_{T_n}(s) \right) \right]^2 \chi_B \, ds \le \frac{4}{\varepsilon^2} Ct \sup_{|x| \le N} \left| \Phi'_{T_n}(x) \right|^2.$$

It is clear that

$$I_2 = \mathsf{P}\left\{ \left| \int\limits_0^t \Phi'_{T_n}\left(\widetilde{\zeta}_{T_n}(s)\right) \widehat{\sigma}_{T_n}\left(\widetilde{\zeta}_{T_n}(s)\right) \chi_{B^c} \, d\widetilde{W}_{T_n}(s) \right| > \frac{\varepsilon}{2} \right\} \le P_N.$$

Therefore, from the previous inequalities passing to the limit as $T_n \to +\infty$ then as $N \to +\infty$ we obtain that right-hand side in (2.17) tends to zero in probability. Consequently, we have the convergence

$$\int\limits_0^t \widehat{\sigma}^2_{T_n}\left(\widetilde{\zeta}_{T_n}(s)\right) ds - \int\limits_0^t \overline{\sigma}^2\left(\widetilde{\zeta}_{T_n}(s)\right) ds \xrightarrow{\widetilde{\mathsf{P}}} 0, \qquad (2.18)$$

as $T_n \to +\infty$.

Taking into account (2.12), we obtain

$$\int\limits_0^t \mathsf{P}\left\{\widetilde{\zeta}(s) = 0\right\} ds = 0.$$

Therefore,

$$\int\limits_0^t \overline{\sigma}^2\left(\widetilde{\zeta}_{T_n}(s)\right) ds \xrightarrow{\widetilde{\mathsf{P}}} \int\limits_0^t \overline{\sigma}^2\left(\widetilde{\zeta}(s)\right) ds,$$

as $T_n \to +\infty$. Taking into account (2.18), we have (2.16). Now, consider the process from relation (2.15):

$$\gamma_{T_n}(t) = \int\limits_0^t \widehat{\sigma}_{T_n}\left(\widetilde{\zeta}_{T_n}(s)\right) d\widetilde{W}_{T_n}(s).$$

For every fixed T_n, $\gamma_{T_n}(t)$ is a martingale with respect to the σ-algebra $\sigma\left\{\gamma_{T_n}(s), \ s \le t\right\}$ and $\gamma_{T_n}(t) \xrightarrow{\widetilde{\mathsf{P}}} \zeta(t)$, as $T_n \to +\infty$. For the quadratic characteristic of this martingale we have

$$\langle \gamma_{T_n} \rangle(t) = \int\limits_0^t \widehat{\sigma}^2_{T_n}\left(\widetilde{\zeta}_{T_n}(s)\right) ds \xrightarrow{\widetilde{\mathsf{P}}} \int\limits_0^t \overline{\sigma}^2\left(\widetilde{\zeta}(s)\right) ds.$$

According to Lemma A.9, the limit process $\widetilde{\zeta}(t)$ is a martingale with respect to the σ-algebra $\sigma\left\{\widetilde{\zeta}(s),\, s \le t\right\}$ whose quadratic characteristic is

$$\langle\widetilde{\zeta}\rangle(t) = \int_0^t \overline{\sigma}^2\left(\widetilde{\zeta}(s)\right) ds.$$

Thus, according to the Doob theorem (see Theorem A.4), there exists a Wiener process $\widehat{W}(t)$ such that

$$\widetilde{\zeta}(t) = \int_0^t \overline{\sigma}\left(\widetilde{\zeta}(s)\right) d\widehat{W}(s). \tag{2.19}$$

Since the sequence $T_n' \to +\infty$ is arbitrary and since the solution $\widetilde{\zeta}(t)$ is strongly unique (see Theorem A.11), then the finite-dimensional distributions of the process $\zeta_T(t)$ converge, as $T \to +\infty$, to the corresponding finite-dimensional distributions of the solution $\zeta(t)$ to Eq. (2.8). According to Theorem A.13, for the weak convergence of the processes $\zeta_T(t)$ to $\zeta(t)$ it is sufficient to prove that for every $N > 0$ and for any $\varepsilon > 0$

$$\lim_{h \to 0} \limsup_{T \to +\infty} \mathsf{P}\left\{\sup_{|t_1-t_2|\le h,\; t_i \le L} |\zeta_T(t_2) - \zeta_T(t_1)| > \varepsilon\right\} = 0. \tag{2.20}$$

In order to prove (2.20), we use the inequalities

$$\sup_{|t_1-t_2|\le h,\; t_i \le L} |\zeta_T(t_2) - \zeta_T(t_1)| \le 2 \sup_{kh \le L} \sup_{kh \le t \le (k+2)h} |\zeta_T(t) - \zeta_T(kh)| \tag{2.21}$$

$$\le 4 \sup_{kh \le L} \sup_{kh \le t \le (k+1)h} |\zeta_T(t) - \zeta_T(kh)|.$$

Therefore, relation (2.20) follows from the inequalities

$$\mathsf{P}\left\{\sup_{|t_1-t_2|\le h;\; t_i \le L} |\zeta_T(t_2) - \zeta_T(t_1)| > \varepsilon\right\}$$

$$\le \mathsf{P}\left\{4 \sup_{kh \le L;\; kh \le t \le (k+1)h} |\zeta_T(t) - \zeta_T(kh)| > \varepsilon\right\}$$

$$\le \sum_{kh \le L} \mathsf{P}\left\{\sup_{kh \le t \le (k+1)h} |\zeta_T(t) - \zeta_T(kh)| > \frac{\varepsilon}{4}\right\}$$

$$\le \sum_{kh \le L} \left(\frac{4}{\varepsilon}\right)^4 \mathsf{E}\left(\int_{kh}^{(k+1)h} \widehat{\sigma}_T\left(\zeta_T(s)\right) dW_T(s)\right)^4 \le C \sum_{kh \le L} h^2.$$

The sufficiency is proved.

Necessity Let the assumptions of Theorem 2.1 hold and let the process ζ_T converge weakly, as $T \to +\infty$, to the solution ζ of Eq. (2.8). Note that the process ζ_T satisfies Eq. (2.14) and $\widehat{\sigma}_T(x) \leq C$. Then for the process

$$\left(\zeta_T(t),\ W_T(t),\ \langle \zeta_T \rangle\,(t)\right),$$

where $\langle \zeta_T \rangle\,(t)$ is the quadratic characteristic of martingale ζ_T, the assumptions of Lemma A.12 (Skorokhod's representation theorem or the Skorokhod's convergent subsequence principle) hold. According to this principle, given an arbitrary sequence $T'_n \to +\infty$, we can choose a subsequence $T_n \to +\infty$, a probability space $(\widetilde{\Omega}, \widetilde{\mathfrak{F}}, \widetilde{\mathsf{P}})$, and a stochastic process

$$\left(\widetilde{\zeta}_{T_n}(t),\ \widetilde{W}_{T_n}(t),\ \widetilde{\langle \zeta_{T_n} \rangle}(t)\right)$$

defined on this space such whose finite-dimensional distributions coincide with those of the process

$$\left(\zeta_{T_n}(t),\ W_{T_n}(t),\ \langle \zeta_{T_n} \rangle(t)\right),$$

and, moreover,

$$\widetilde{\zeta}_{T_n}(t) \xrightarrow{\ \widetilde{\mathsf{P}}\ } \widetilde{\zeta}(t),\quad \widetilde{W}_{T_n}(t) \xrightarrow{\ \widetilde{\mathsf{P}}\ } \widetilde{W}(t),\quad \widetilde{\langle \zeta_{T_n} \rangle}(t) \xrightarrow{\ \widetilde{\mathsf{P}}\ } \beta(t),$$

as $T_n \to +\infty$, where $\widetilde{\zeta}(t),\ \widetilde{W}(t),\ \beta(t)$ are some stochastic processes.

According to Lemma A.13, relation (2.15) holds for the processes $\widetilde{\zeta}_{T_n}(t)$ and $\widetilde{W}_{T_n}(t)$, as well. Therefore, $\widetilde{\zeta}_{T_n}(t)$ is a martingale for every fixed T_n with respect to the σ-algebra $\sigma\left\{\widetilde{\zeta}_{T_n}(s),\, s \leq t\right\}$. For the quadratic characteristic of this martingale we have for some $\beta(t)$ the convergence

$$\widetilde{\langle \zeta_{T_n} \rangle}(t) = \int\limits_0^t \widehat{\sigma}_{T_n}^2\left(\widetilde{\zeta}_{T_n}(s)\right) ds \xrightarrow{\ \widetilde{\mathsf{P}}\ } \beta(t),$$

as $T_n \to +\infty$. Note that $\widetilde{\zeta}_{T_n}(0) \to 0$, as $T_n \to +\infty$, with probability 1.

According to Lemma A.9, we obtain that $\widetilde{\zeta}$ is a martingale with respect to the σ-algebra $\sigma\left\{\widetilde{\zeta}(s),\, s \leq t\right\}$ with the quadratic characteristic $\langle \widetilde{\zeta} \rangle(t) = \beta(t)$. Since the finite-dimensional distributions of the processes $\widetilde{\zeta}(t)$ and $\zeta(t)$ coincide, according to Eq. (2.8) and the boundedness of the coefficient σ, we obtain that

$$\langle \zeta \rangle\,(t) = \int\limits_0^t \overline{\sigma}^2\,(\zeta(s))\ ds.$$

Consequently

$$\beta(t) = \int_0^t \overline{\sigma}^2 \left(\xi(s) \right) ds.$$

Therefore,

$$\int_0^t \widehat{\sigma}_{T_n}^2 \left(\xi_{T_n}(s) \right) ds \xrightarrow{\widetilde{P}} \int_0^t \overline{\sigma}^2 \left(\xi(s) \right) ds,$$

as $T_n \to +\infty$.

Taking into account (2.12), we obtain the following convergence

$$\int_0^t \widehat{\sigma}_{T_n}^2 \left(\xi_{T_n}(s) \right) ds - \int_0^t \overline{\sigma}^2 \left(\xi_{T_n}(s) \right) ds \xrightarrow{\widetilde{P}} 0$$

for all $t \geq 0$. Using the Itô formula, we have

$$\int_0^t \widehat{\sigma}_{T_n}^2 \left(\xi_{T_n}(s) \right) ds - \int_0^t \overline{\sigma}^2 \left(\xi_{T_n}(s) \right) ds$$

$$= \Phi_{T_n} \left(\xi_{T_n}(t) \right) - \Phi_{T_n} \left(\xi_{T_n}(0) \right) - \int_0^t \Phi'_{T_n} \left(\xi_{T_n}(s) \right) d\xi_{T_n}(s),$$

where

$$\Phi_{T_n}(x) = 2 \int_0^x \left[u - \overline{\sigma}^2(u) Q_{T_n}(u) \right] du.$$

Therefore

$$\Phi_{T_n} \left(\xi_{T_n}(t) \right) - \Phi_{T_n} \left(\xi_{T_n}(0) \right) - \int_0^t \Phi'_{T_n} \left(\xi_{T_n}(s) \right) d\xi_{T_n}(s) \xrightarrow{\widetilde{P}} 0, \qquad (2.22)$$

as $T_n \to +\infty$, for every $t \geq 0$.

The functions $Q_{T_n}(x)$ are increasing in x for every T_n. Using the assumption (2.11) and the fact that $\varphi(x)$ is the inverse function to $f(x)$, we obtain

$$
\left| Q_{T_n}(x) \right| = \left| \frac{1}{\sqrt{T_n}} \int\limits_0^{\varphi(x\sqrt{T_n})} \frac{du}{f'(u)\sigma^2(u)} \right|
$$

$$
= |x| \left| \frac{1}{f\left(\varphi\left(x\sqrt{T_n}\right)\right)} \int\limits_0^{\varphi(x\sqrt{T_n})} \frac{du}{f'(u)\,\sigma^2(u)} \right| \leq C\,|x|.
$$

Therefore, there exist a subsequence $\widetilde{T}_n \to +\infty$ of the sequence T_n and an increasing function $Q(x)$ such that $Q_{\widetilde{T}_n}(x) \to Q(x)$, as $\widetilde{T}_n \to +\infty$, at each point of continuity of the function $Q(x)$.

Let us pass to the limit, as $\widetilde{T}_n \to +\infty$, in each term of the left-hand part in (2.22). In order to do this, we introduce the function

$$
\widehat{\Phi}(x) = 2 \int\limits_0^x \left[u - \overline{\sigma}^2(u)\, Q(u) \right] du,
$$

and obtain for any $\widetilde{T}_{n_0}$ that

$$
\int\limits_0^t \Phi'_{\widetilde{T}_n}\left(\zeta_{\widetilde{T}_n}(s)\right) d\zeta_{\widetilde{T}_n}(s) - \int\limits_0^t \widehat{\Phi}'\left(\widetilde{\zeta}(s)\right) d\widetilde{\zeta}(s)
$$

$$
= \left(\int\limits_0^t \Phi'_{\widetilde{T}_{n_0}}\left(\zeta_{\widetilde{T}_n}(s)\right) d\zeta_{\widetilde{T}_n}(s) - \int\limits_0^t \Phi'_{\widetilde{T}_{n_0}}\left(\widetilde{\zeta}(s)\right) d\widetilde{\zeta}(s) \right)
$$

$$
+ \left(\int\limits_0^t \Phi'_{\widetilde{T}_{n_0}}\left(\widetilde{\zeta}(s)\right) d\widetilde{\zeta}(s)) - \int\limits_0^t \widehat{\Phi}'\left(\widetilde{\zeta}(s)\right) d\widetilde{\zeta}(s) \right)
$$

$$
+ \left(\int\limits_0^t \Phi'_{\widetilde{T}_n}\left(\zeta_{\widetilde{T}_n}(s)\right) d\zeta_{\widetilde{T}_n}(s) - \int\limits_0^t \Phi'_{\widetilde{T}_{n_0}}\left(\zeta_{\widetilde{T}_n}(s)\right) d\zeta_{\widetilde{T}_n}(s) \right)
$$

$$
= I_1 + I_2 + I_3.
$$

It is clear that the function $\Phi_{\widetilde{T}_{n_0}}(x)$ is continuous in x for each $\widetilde{T}_{n_0}$. According to Lemma A.7, $I_1 \xrightarrow{\widetilde{\mathsf{P}}} 0$, as $\widetilde{T}_n \to +\infty$.

Let τ_N be the moment of the first exit of the process $\widetilde{\zeta}_{\widetilde{T}_n}(t)$ from interval $(-N, N)$.

According to Lemma A.10, we have

$$\mathsf{P}\{|I_3| > \varepsilon\} \leq \mathsf{P}\left\{\sup_{0 \leq s \leq t} \left|\widetilde{\zeta}_{\widetilde{T}_n}(s)\right| > N\right\}$$

$$+ \frac{4}{\varepsilon^2} \mathsf{E}\left(\int_0^{t \wedge \tau_N} \left[\Phi'_{\widetilde{T}_n}\left(\widetilde{\zeta}_{T_n}(s)\right) - \Phi'_{T_{n_0}}\left(\widetilde{\zeta}_{\widetilde{T}_n}(s)\right)\right] d\widetilde{\zeta}_{\widetilde{T}_n}(s)\right)^2$$

$$\leq \mathsf{P}\left\{\sup_{0 \leq s \leq t} \left|\widetilde{\zeta}_{\widetilde{T}_n}(s)\right| > N\right\} + \frac{4}{\varepsilon^2} C \int_{-N}^{N} \left|\Phi'_{\widetilde{T}_n}(x) - \Phi'_{\widetilde{T}_{n_0}}(x)\right|^2 dx.$$

Similarly,

$$\mathsf{P}\{|I_2| > \varepsilon\} \leq \mathsf{P}\left\{\sup_{0 \leq s \leq t} \left|\widetilde{\zeta}(s)\right| > N\right\} + \frac{4}{\varepsilon^2} C \int_{-N}^{N} \left|\Phi'_{\widetilde{T}_{n_0}}(x) - \widehat{\Phi}'(x)\right|^2 dx$$

for any $\varepsilon > 0$, $L > 0$ and $N > 0$. Therefore, $I_2 \xrightarrow{\widetilde{\mathsf{P}}} 0$, $I_3 \xrightarrow{\widetilde{\mathsf{P}}} 0$, as $\widetilde{T}_n \to +\infty$, $\widetilde{T}_{n_0} \to +\infty$.

Consequently,

$$\int_0^t \Phi'_{\widetilde{T}_n}\left(\widetilde{\zeta}_{\widetilde{T}_n}(s)\right) d\widetilde{\zeta}_{\widetilde{T}_n}(s) \xrightarrow{\widetilde{\mathsf{P}}} \int_0^t \widehat{\Phi}'\left(\widetilde{\zeta}(s)\right) d\widetilde{\zeta}(s),$$

as $\widetilde{T}_n \to +\infty$. Furthermore, for any $N > 0$

$$\sup_{|x| \leq N} \left|\Phi_{\widetilde{T}_n}(x) - \widehat{\Phi}(x)\right| \leq C \int_{-N}^{N} \left|Q_{\widetilde{T}_n}(u) - Q(u)\right| du \to 0,$$

as $\widetilde{T}_n \to +\infty$. Thus,

$$\Phi_{\widetilde{T}_n}\left(\widetilde{\zeta}_{\widetilde{T}_n}(t)\right) \xrightarrow{\widetilde{\mathsf{P}}} \widehat{\Phi}\left(\widetilde{\zeta}(t)\right),$$

as $\widetilde{T}_n \to +\infty$. Taking into account (2.22), we have the equality

$$\widehat{\Phi}\left(\widetilde{\zeta}(t)\right) - \int_0^t \widehat{\Phi}'\left(\widetilde{\zeta}(s)\right) d\widetilde{\zeta}(s) = 0 \tag{2.23}$$

with probability 1 for all $t \geq 0$.

Using (2.23) and Lemma A.12, we have that $\widehat{\Phi}'(x) = b$ for all x, where b is some constant. Using condition $\widehat{\Phi}'(0) = 0$, we obtain $\widehat{\Phi}'(x) = 0$ for all x. Therefore,

$$Q(x) = \frac{x}{\overline{\sigma}^2(x)}$$

for all x.

Taking into account the relation $Q_{\widetilde{T}_n}(x) \to Q(x)$, as $\widetilde{T}_n \to +\infty$, we get that for $x \neq 0$

$$\frac{1}{f\left(\varphi\left(x\sqrt{\widetilde{T}_n}\right)\right)} \int_0^{\varphi\left(x\sqrt{\widetilde{T}_n}\right)} \frac{du}{f'(u)\,\sigma^2(u)} \to \frac{1}{\overline{\sigma}^2(x)},$$

as $\widetilde{T}_n \to +\infty$.

Since the sequence $\widetilde{T}_n \to +\infty$ is arbitrary, we have relation (2.13). Theorem 2.1 is proved. □

Corollary 2.1 *Since the processes ζ_T and W_T satisfy the assumptions of Lemma A.13 and Theorem A.12, in order to study the weak convergence of this processes, as $T \to +\infty$, without loss of generality, we can assume that for an arbitrary sequence $T_n \to +\infty$ there exist some processes ζ and W such that $\zeta_{T_n}(t) \xrightarrow{P} \zeta(t)$, $W_{T_n}(t) \xrightarrow{P} W(t)$, as $T_n \to +\infty$.*

The considerations of Corollary 2.1 are used for simplification of the proof of some theorems, for example, the proofs of Theorems 4.6 and 4.8 can be simplified.

We proceed with some applications of Theorem 2.1. The next remark presents important examples of coefficients a and σ, satisfying conditions (2.4), (2.11), and (2.13).

Remark 2.5 The conditions of Theorem 2.1 are satisfied in the following cases:

(1) if there exist a derivative $\sigma'(x)$ of the function $\sigma(x)$ and the integrals

$$\lambda_1 := \int_0^{+\infty} \left[\frac{a(x)}{\sigma^2(x)} - \frac{1}{2}\frac{\sigma'(x)}{\sigma(x)} \right] dx, \quad \lambda_2 := \int_0^{-\infty} \left[\frac{a(x)}{\sigma^2(x)} - \frac{1}{2}\frac{\sigma'(x)}{\sigma(x)} \right] dx$$

with $\sigma_1 = \sigma(0)e^{-2\lambda_1}$ and $\sigma_2 = \sigma(0)e^{-2\lambda_2}$, where

$$\int_{\mathbb{R}} \left[\frac{a(x)}{\sigma^2(x)} - \frac{1}{2} \frac{\sigma'(x)}{\sigma(x)} \right] dx = \lambda_1 - \lambda_2 = \lambda;$$

(2) if the following spatial averaging holds:

$$\frac{1}{x} \int_0^x \exp \left\{ -2 \int_0^u \frac{a(v)}{\sigma^2(v)} dv \right\} du \to \begin{cases} \lambda_1, & x \to +\infty, \\ \lambda_2, & x \to -\infty, \end{cases}$$

$$\frac{1}{x} \int_0^x \exp \left\{ 2 \int_0^u \frac{a(v)}{\sigma^2(v)} dv \right\} \frac{du}{\sigma^2(u)} \to \begin{cases} \lambda_3, & x \to +\infty, \\ \lambda_4, & x \to -\infty, \end{cases}$$

$$0 < \lambda_i < +\infty$$

with $\sigma_1 = \sqrt{\frac{\lambda_1}{\lambda_3}}$ and $\sigma_2 = \sqrt{\frac{\lambda_2}{\lambda_4}}$.

Note that case (2) may include, in particular, vibrational coefficients, for example $a(x) = \sin x$, $\sigma(x) = 1$, etc.

Now, our goal is to study the asymptotic behavior, as $t \to +\infty$, of the distribution of the solution ξ to Eq. (2.1). Note that if we know the behavior of the distribution of $f(\xi(t))$ for regularly varying (at infinity) functions $f(x)$ with index $\alpha > 0$ (see Definition 4.1), we can find the behavior of the distribution of the solution ξ itself.

Lemma 2.3 *Let the conditions of Theorem 2.1 be satisfied and let the function $f(x)$ be such that*

$$\lim_{k \to +\infty} \frac{f(kx)}{f(k)} = b(x) = \begin{cases} x^\alpha, & x > 0, \\ -c_0 |x|^\alpha, & x < 0, \end{cases} \tag{2.24}$$

where $\alpha > 0$ and $c_0 > 0$. Also, let the normalizing factor $B = B(T) > 0$ be a solution to the equation $\sqrt{T} = f(B(T))$.
   Then the finite-dimensional distributions of the process

$$\xi_T(t) = \frac{\xi(tT)}{B(T)}$$

converge, as $T \to +\infty$, to the corresponding finite-dimensional distributions of a homogeneous Markov process with the transition density

$$\rho_1(t, x, y) = \rho(t, b(x), b(y)) b'(y), \tag{2.25}$$

where the function $\rho(t, x, y)$ is defined by (2.10).

Proof Consider the normalized random process $\xi_T(t) = \frac{\xi(tT)}{B(T)}$, $t > 0$, where $B(T)$ is the solution to the equation $\sqrt{T} = f(B(T))$.

Note that for a monotonically increasing function $f(x)$ there exists the inverse function $\varphi(x)$ and the normalizing factor has the form $B(T) = \varphi\left(\sqrt{T}\right)$.

So,

$$\xi_T(t) = \frac{\xi(tT)}{\varphi\left(\sqrt{T}\right)} = \frac{\varphi\left(f\left(\xi(tT)\right)\right)}{\varphi\left(\sqrt{T}\right)} = \frac{\varphi\left(\frac{f(\xi(tT))}{\sqrt{T}} \cdot \sqrt{T}\right)}{\varphi\left(\sqrt{T}\right)} = \frac{\varphi\left(\zeta_T(t)\sqrt{T}\right)}{\varphi\left(\sqrt{T}\right)}.$$

Taking into account the weak equivalence of the processes ζ_{T_n} and $\tilde{\zeta}_{T_n}$ (see Theorem 2.1), let us establish that the processes

$$\tilde{\xi}_{T_n}(t) = \frac{\varphi\left(\tilde{\zeta}_{T_n}(t)\sqrt{T_n}\right)}{\varphi\left(\sqrt{T_n}\right)}$$

and $\xi_{T_n}(t)$ are weakly equivalent for all T_n.

Indeed,

$$P\left\{\xi_{T_n}(t) < x\right\} = P\left\{\frac{\varphi\left(\zeta_{T_n}(t)\sqrt{T_n}\right)}{\varphi\left(\sqrt{T_n}\right)} < x\right\} = P\left\{\varphi\left(\zeta_{T_n}(t)\sqrt{T_n}\right) < x\varphi\left(\sqrt{T_n}\right)\right\}$$

$$= P\left\{\zeta_{T_n}(t)\sqrt{T_n} < f\left(x\varphi\left(\sqrt{T_n}\right)\right)\right\} = P\left\{\zeta_{T_n}(t) < \frac{f\left(x\varphi\left(\sqrt{T_n}\right)\right)}{\sqrt{T_n}}\right\},$$

and in a similar way we have

$$\tilde{P}\left\{\tilde{\xi}_{T_n}(t) < x\right\} = \tilde{P}\left\{\tilde{\zeta}_{T_n}(t) < \frac{f\left(x\varphi\left(\sqrt{T_n}\right)\right)}{\sqrt{T_n}}\right\}.$$

Consequently,

$$P\left\{\xi_{T_n}(t) < x\right\} = \tilde{P}\left\{\tilde{\xi}_{T_n}(t) < x\right\}.$$

Similarly we can prove that for any $t_1, t_2, \ldots, t_l$ and $x_1, x_2, \ldots, x_l$

$$P\left\{\xi_{T_n}(t_1) < x_1, \xi_{T_n}(t_2) < x_2, \ldots, \xi_{T_n}(t_l) < x_l\right\}$$

$$= \tilde{P}\left\{\tilde{\xi}_{T_n}(t_1) < x_1, \tilde{\xi}_{T_n}(t_2) < x_2, \ldots, \tilde{\xi}_{T_n}(t_l) < x_l\right\}.$$

From the proof of Theorem 2.1 we can get that $\tilde{\zeta}_{T_n}(t) \xrightarrow{\tilde{P}} \tilde{\zeta}(t)$, as $T_n \to +\infty$.
It is clear that for regularly varying at infinity function $\varphi(x)$ we have the relation

$$\lim_{k \to +\infty} \frac{\varphi(kx)}{\varphi(k)} = b_1(x) = \begin{cases} x^{\frac{1}{\alpha}}, & x > 0, \\ -\left(\frac{1}{c_0}|x|\right)^{\frac{1}{\alpha}}, & x < 0, \end{cases}$$

where $b_1(x)$ is the inverse function to the function $b(x)$.

Using the properties from [25] of regularly varying at infinity functions (see Lemma A.17), we can get that

$$\sup_{0 < \delta \leq |x| \leq N} \left| \frac{\varphi\left(|x|\sqrt{T}\right)}{\varphi\left(\sqrt{T}\right)} - |x|^{\alpha} \right| \to 0,$$

as $T \to +\infty$, for all $0 < \delta < N < +\infty$. Therefore

$$\tilde{\xi}_{T_n}(t) \xrightarrow{\tilde{P}} b_1\left(\tilde{\zeta}(t)\right),$$

as $T_n \to +\infty$.

Now, the statement of lemma follows from the proof of Theorem 2.1 and the monotonicity of the function $b_1(x)$. $\qquad\square$

Remark 2.6 If conditions (2.24) are fulfilled with $\alpha = 1$, then the finite-dimensional distributions of the process $\xi_T(t) = \xi(tT)B^{-1}(T)$ converge, as $T \to +\infty$, to the corresponding finite-dimensional distributions of the process $\widehat{\xi}(t)$, which is defined by (2.9) with $c_1 = 1$, $c_2 = \frac{1}{c_0}$.

It is easy to obtain that for the process $\widehat{\xi}(t)$ of the form (2.9), there exist Kolmogorov's local characteristics in the generalized sense: the diffusion coefficient $\overline{\sigma}^2(x)\overline{c}^{-2}(x)$ and the drift coefficient $c\delta(x)$, where $\delta(x)$ is Dirac's delta function,

$$c = \frac{1}{2}\frac{\sigma_1}{c_1} \cdot \frac{\sigma_2}{c_2}\left(\frac{\sigma_1}{c_1} + \frac{\sigma_2}{c_2}\right)\frac{c_2 - c_1}{\sigma_2 + \sigma_1}. \tag{2.26}$$

In particular, if option (1) from Remark 2.5 holds with $\sigma(x) = 1$, then the stochastic process $\xi_T(t) = \xi(tT)T^{-\frac{1}{2}}$ converges weakly, as $T \to +\infty$, to the process $\widehat{\xi}(t)$ with $c_1 = \sigma_1$ and $c_2 = \sigma_2$, and

$$c = \frac{\sigma_2 - \sigma_1}{\sigma_2 + \sigma_1} = \frac{e^{-2\lambda_2} - e^{-2\lambda_1}}{e^{-2\lambda_2} + e^{-2\lambda_1}} = \frac{e^{-2[\lambda_2 - \lambda_1]} - 1}{e^{-2[\lambda_2 - \lambda_1]} + 1} = \frac{e^{2\lambda} - 1}{e^{2\lambda} + 1} = \frac{e^{\lambda} - e^{-\lambda}}{e^{\lambda} + e^{-\lambda}} = \tanh \lambda,$$

where

$$\lambda = \int_{\mathbb{R}} a(x)\,dx = \int_0^{+\infty} a(x)\,dx + \int_{-\infty}^0 a(x)\,dx = \int_0^{+\infty} a(x)\,dx - \int_0^{-\infty} a(x)\,dx = \lambda_1 - \lambda_2.$$

If option (2) from Remark 2.5 holds with $\sigma(x) = 1$, then the stochastic process $\xi_T(t) = \xi(tT)T^{-\frac{1}{2}}$ converges weakly, as $T \to +\infty$, to the process $\widehat{\xi}(t)$ with $c_1 = \lambda_1$, $c_2 = \lambda_2$, $\sigma_1 = \sqrt{\frac{\lambda_1}{\lambda_3}}$, $\sigma_2 = \sqrt{\frac{\lambda_2}{\lambda_4}}$. In particular, for $\lambda_1 = \lambda_2$ and $\lambda_3 = \lambda_4$ the limit process is $\widehat{\xi}(t) = \frac{1}{\sqrt{\lambda_1 \lambda_3}} W(t)$.

2.3 Necessary and Sufficient Conditions for the Weak Convergence of Solutions of SDEs to a Process of Skew Brownian Motion Type

For a certain class of Eq. (2.1), condition (2.13) is necessary and sufficient for the weak convergence of $\xi_T(t) = \xi(tT)T^{-\frac{1}{2}}$ to the process $\widehat{\xi}(t)$ of the form (2.9).

Theorem 2.2 *Let ξ be a solution to Eq. (2.1), and let coefficients satisfy the following conditions: $0 < \sigma(x) < C$,*

$$\left| \int_0^x \frac{a(u)}{\sigma^2(u)}\,du \right| < C, \quad and \quad \lim_{|x| \to +\infty} \left(\frac{1}{f(x)} \int_0^x \frac{du}{f'(u)\sigma^2(u)} - \bar{b}^2(x) \right) = 0,$$

where

$$\bar{b}(x) = \begin{cases} b_1, & x \geq 0, \\ b_2, & x < 0, \end{cases} \quad 0 < b_i < +\infty.$$

Then the stochastic process $\xi_T(t) = \xi(tT)T^{-\frac{1}{2}}$ converges weakly, as $T \to +\infty$, to the process $\widehat{\xi}(t)$, defined by relation (2.9), if and only if

$$\lim_{|x| \to +\infty} \left(\frac{f(x)}{x} - k\bar{c}(x) \right) = 0, \quad \bar{b}(x) = \frac{1}{k\bar{\sigma}(x)},$$

where k is some positive constant.

Proof *Sufficiency* It is clear that in this case the conditions of Theorem 2.1 are fulfilled.

In addition, we have (2.24) with $\alpha = 1$ and $c_0 = \frac{c_2}{c_1}$. Herewith $B(T) \sim \frac{1}{kc_1}\sqrt{T}$, as $T \to +\infty$. Therefore, using Remark 2.5, we have that the finite-dimensional distributions of the process $\xi_T(t) = \xi(tT)T^{-\frac{1}{2}}$ converge to the corresponding finite-dimensional distributions of the process $\widehat{\xi}(t) = \frac{1}{kc_1}b_1\left(\widetilde{\zeta}(t)\right)$, where

$$b_1(x) = \begin{cases} x, & x \geq 0, \\ \frac{c_1}{c_2}x, & x < 0, \end{cases}$$

and $\widetilde{\zeta}(t)$ is the solution to the equation

$$\widetilde{\zeta}(t) = \int\limits_0^t k\overline{\sigma}\left(\widetilde{\zeta}(s)\right)dW(s), \quad \overline{\sigma}(x) = \begin{cases} \sigma_1, & x \geq 0, \\ \sigma_2, & x < 0. \end{cases}$$

Since $\overline{\sigma}(x) = \overline{\sigma}\left(\frac{x}{k}\right)$, the process $\zeta(t) = k^{-1}\widetilde{\zeta}(t)$ is the solution to Eq. (2.8). Thus,

$$\widehat{\xi}(t) = \frac{1}{kc_1}b_1\left(\widetilde{\zeta}(t)\right) = l\left(\zeta(t)\right),$$

where

$$l(x) = \begin{cases} \frac{1}{c_1}x, & x \geq 0, \\ \frac{1}{c_2}x, & x < 0. \end{cases}$$

To complete the proof of sufficiency it remains to show that (2.20) holds for the process ξ_T. In fact, using the inequality $f'(x) \geq \delta > 0$, we have

$$|\xi_T(t_1) - \xi_T(t_2)| = \frac{1}{\sqrt{T}}\left|\varphi\left(\zeta_T(t_1)\sqrt{T}\right) - \varphi\left(\zeta_T(t_2)\sqrt{T}\right)\right|$$

$$\leq C\,|\zeta_T(t_1) - \zeta_T(t_2)|,$$

where $\varphi(x)$ is the inverse function to $f(x)$. This inequality yields (2.20) for the process ξ_T for every $N > 0$ with any $\varepsilon > 0$.

Necessity Let the assumptions of Theorem 2.2 be fulfilled and let the process $\xi_T(t) = \xi(tT)T^{-\frac{1}{2}}$ converge weakly, as $T \to +\infty$, to the solution $\widehat{\xi}(t)$ of Eq. (2.9).

Note that $\xi_T(t) = \varphi_T\left(\zeta_T(t)\right)$, where $\varphi_T(x) = T^{-\frac{1}{2}}\varphi(x\sqrt{T})$, and the process $\zeta_T(t)$ is the solution to Eq. (2.14). According to Theorem A.12, given an arbitrary sequence $T''_n \to +\infty$, we can choose a subsequence $T'_n \to +\infty$, a probability space $(\widetilde{\Omega}, \widetilde{\mathfrak{F}}, \widetilde{\mathsf{P}})$, and a stochastic process $\widetilde{\zeta}_{T'_n}(t)$, defined on this space, which is weakly equivalent to the process $\zeta_{T'_n}(t)$, herewith $\widetilde{\zeta}_{T'_n}(t) \xrightarrow{\widetilde{\mathsf{P}}} \widetilde{\zeta}(t)$, as $T'_n \to +\infty$.

In addition, the function $\varphi_T(x)$ is monotonically increasing in x for each T and $|\varphi_T(x)| \leq C|x|$. Since $\varphi_T'(x) = \frac{1}{f'(\varphi_T(x))}$, we have that $0 < \delta \leq \varphi_T'(x) \leq C$. Therefore, there exists a subsequence $T_n \to +\infty$ of the sequence T_n' and a monotonically increasing continuous function $\widetilde{\varphi}(x)$ such that $\varphi_{T_n}(x) \to \widetilde{\varphi}(x)$ at every point x, as $T_n \to +\infty$. Thus,

$$\widetilde{\xi}_{T_n}(t) = \varphi_{T_n}\left(\widetilde{\zeta}_{T_n}(t)\right) \xrightarrow{\widetilde{P}} \widetilde{\varphi}\left(\widetilde{\zeta}(t)\right)$$

at every point $t > 0$, as $T_n \to +\infty$.

It follows from the proof of Theorem 2.1, that the process $\widetilde{\zeta}(t)$ is the solution to Eq. (2.19) with $\overline{\sigma}(x) = \frac{1}{\overline{b}(x)}$. Note that the processes $\xi_{T_n}(t)$ and $\widetilde{\xi}_{T_n}(t)$ are weakly equivalent and the process $\xi_{T_n}(t)$ converges weakly, as $T_n \to +\infty$, to the process $l(\zeta(t))$, defined by (2.9). Therefore $\widetilde{\varphi}\left(\widetilde{\zeta}(t)\right) = l\left(\zeta(t)\right)$, where $\zeta(t)$ is the solution to equation

$$\zeta(t) = \int_0^t \overline{\sigma}\left(\zeta(s)\right) d\widehat{W}(s).$$

Using the equalities $\zeta(t) = l^{-1}\left(\widetilde{\varphi}\left(\widetilde{\zeta}(t)\right)\right)$, $\overline{\sigma}\left(\zeta(t)\right) = \overline{\sigma}\left(\widetilde{\zeta}(t)\right)$ and Eq. (2.19), we obtain

$$l^{-1}\left(\widetilde{\varphi}\left(\widetilde{\zeta}(t)\right)\right) = \int_0^t \overline{\sigma}\left(\widetilde{\zeta}(s)\right) \overline{b}\left(\widetilde{\zeta}(s)\right) d\widetilde{\zeta}(s)$$

with probability 1 for all $t > 0$. According to Lemma A.12, we have the equalities

$$\frac{l^{-1}(\widetilde{\varphi}(x))}{x} = \overline{\sigma}(x)\,\overline{b}(x) = k_0$$

for all x, where k_0 is some constant. From here we have that

$$\overline{b}(x) = \frac{k_0}{\overline{\sigma}(x)}, \qquad \widetilde{\varphi}(x) = x\frac{k_0}{\overline{c}(x)}.$$

Thus,

$$f_{T_n}(x) \to \widetilde{\varphi}^{-1}(x) = x\frac{\overline{c}(x)}{k_0},$$

as $T_n \to +\infty$. Since

$$f_{T_n}(x) = \frac{f\left(x\sqrt{T_n}\right)}{\sqrt{T_n}} = x\frac{f\left(x\sqrt{T_n}\right)}{x\sqrt{T_n}},$$

and the sequence $T_n \to +\infty$ is arbitrary, we have the convergence

$$\frac{f(x)}{x} - \frac{\overline{c}(x)}{k_0} \to 0,$$

as $|x| \to +\infty$.

Theorem is proved with $k = \frac{1}{k_0}$.   □

Theorem 2.3 *Let ξ be a solution to Eq. (2.1) and let*

$$0 < \delta \le \sigma(x) \le C, \quad \left| \int\limits_0^x \frac{a(u)}{\sigma^2(u)} du \right| \le C.$$

The stochastic process $\xi_T(t) = \xi(tT)T^{-\frac{1}{2}}$ converges weakly, as $T \to +\infty$, to the process $\widehat{\xi}(t)$, defined by relation (2.9), if and only if

$$\frac{f(x)}{x} - k\overline{c}(x) \to 0, \quad \frac{1}{f(x)} \int\limits_0^x \frac{du}{f'(u)\sigma^2(u)} - \frac{1}{k^2 \overline{\sigma}^2(x)} \to 0$$

as $|x| \to +\infty$, where k is some positive constant.

Proof *Sufficiency* Sufficiency of the conditions of this theorem follows from Theorem 2.2.

Necessity Let the assumptions of Theorem 2.2 be fulfilled. It is clear that

$$0 < \delta \le f'(x) \le C \quad \text{and} \quad 0 < \delta \le f'(x)\sigma(x) \le C$$

for some constants $\delta > 0$ and $C > 0$. Therefore, the families of monotonically increasing functions $f_T(x)$ and $Q_T(x)$ are compact. Consequently, given an arbitrary sequence $T'_n \to +\infty$ we can choose a subsequence $T_n \to +\infty$ and monotonically increasing functions $\widehat{f}(x)$ and $Q(x)$ such that

$$f_{T_n}(x) \to \widehat{f}(x), \quad Q_{T_n}(x) \to Q(x),$$

as $T_n \to +\infty$, at the points of continuity of the limit functions. The derivatives $\widehat{f}'(x)$ and $Q'(x)$, which exist a.e. with respect to the Lebesgue measure, satisfy the inequalities $0 < \delta \le \widehat{f}'(x) \le C, 0 < \delta \le Q'(x) \le C$. In addition, in this case the conditions of Theorem 2.1 are fulfilled. Therefore, without loss of generality, we will assume that for the corresponding processes $\widetilde{\zeta}_{T_n}(t)$ and $\widetilde{W}_{T_n}(t)$ in Theorem 2.1, we have the equality (2.15) and the convergence

$$\widetilde{\zeta}_{T_n}(t) \xrightarrow{\widetilde{P}} \widetilde{\zeta}(t), \quad \widetilde{W}_{T_n}(t) \xrightarrow{\widetilde{P}} \widetilde{W}(t),$$

as $T_n \to +\infty$.

Next, consider the processes

$$\eta_{T_n}(t) = \int\limits_0^t \sqrt{Q'\left(\tilde{\zeta}_{T_n}(s)\right)} \, d\tilde{\zeta}_{T_n}(s) . \qquad (2.27)$$

Since $Q'(x)$ exists everywhere except the set Λ of zero Lebesgue measure, the integral on the right-hand side in (2.27) is correctly defined. According to Lemma A.11, we have that

$$\int\limits_0^t \tilde{\mathsf{P}}\left\{\tilde{\zeta}_{T_n}(s) \in \Lambda\right\} ds = 0$$

for all $t \geq 0$.

Let us prove that the quadratic characteristic $\langle \eta_{T_n} \rangle(t)$ of the family of martingales $\eta_{T_n}(t)$ converges in probability to t, as $T_n \to +\infty$. In order to do this, we introduce the functions

$$\Phi_{T_n}(x) = 2 \int\limits_0^x \left[Q(u) - Q_{T_n}(u) \right] du,$$

and, using the Itô formula, obtain that

$$\langle \eta_{T_n} \rangle(t) - t = \Phi_{T_n}(\tilde{\zeta}_{T_n}(t)) - \Phi_{T_n}(\tilde{\zeta}_{T_n}(0)) - \int\limits_0^t \Phi'_{T_n}(\tilde{\zeta}_{T_n}(s)) \, d\tilde{\zeta}_{T_n}(s). \quad (2.28)$$

Further, for any $N > 0$

$$\left| \Phi_{T_n}(x) \right| \chi_{|x| \leq N} \leq 2 \int\limits_{-N}^N \left| Q(u) - Q_{T_n}(u) \right| du \to 0,$$

as $T_n \to +\infty$, and according to Lemma A.10

$$\mathsf{E} \int\limits_0^t \left[\Phi'_{T_n}(\tilde{\zeta}_{T_n}(s)) \, \hat{\sigma}_{T_n}(\tilde{\zeta}_{T_n}(s)) \right]^2 \chi_{|\tilde{\zeta}_{T_n}(s)| \leq N} \, ds \leq C_N \int\limits_{-N}^N \left| Q(u) - Q_{T_n}(u) \right|^2 du.$$

From the above we have the obvious inequalities

$$\widetilde{P}\left\{\left|\varPhi_{T_n}\left(\widetilde{\zeta}_{T_n}(s)\right) - \varPhi_{T_n}\left(\widetilde{\zeta}_{T_n}(0)\right)\right| > \varepsilon\right\}$$

$$\leq \widetilde{P}\left\{\sup_{0\leq s \leq t} \left|\widetilde{\zeta}_{T_n}(s)\right| > N\right\} + \frac{4}{\varepsilon}\int_{-N}^{N}\left|G(u) - G_{T_n}(u)\right| du,$$

$$\widetilde{P}\left\{\left|\int_0^t \varPhi'_{T_n}\left(\widetilde{\zeta}_{T_n}(s)\right) d\widetilde{\zeta}_{T_n}(s)\right| > \varepsilon\right\} \leq \widetilde{P}\left\{\sup_{0\leq s \leq t}\left|\widetilde{\zeta}_{T_n}(s)\right| > N\right\}$$

$$+ \frac{4}{\varepsilon^2}\mathsf{E}\int_0^t \left[\varPhi'_{T_n}\left(\widetilde{\zeta}_{T_n}(s)\right)\widehat{\sigma}_{T_n}\left(\widetilde{\zeta}_{T_n}(s)\right)\right]^2 \chi_{|\widetilde{\zeta}_{T_n}(s)|\leq N}\, ds.$$

It follows from (2.28) that $\langle \eta_{T_n}\rangle(t) \xrightarrow{\widetilde{P}} t$, as $T_n \to +\infty$. Consequently, the process $\widehat{W}(t)$, that is the limit in probability of the sequence $\eta_{T_n}(t)$, is a Wiener process, and, according to (2.27), we have the equality

$$\widehat{W}(t) = \int_0^t \sqrt{Q\left(\widetilde{\zeta}(s)\right)}\, d\widetilde{\zeta}(s) \tag{2.29}$$

with probability 1 for all $t > 0$.

Consider the process $\widetilde{\xi}_{T_n}(t) = \varphi_{T_n}\left(\widetilde{\zeta}_{T_n}(t)\right)$, where $\varphi_{T_n}(x)$ is the inverse function to $f_{T_n}(x)$. It is clear that $\widetilde{\xi}_{T_n}(t)$ converges in probability, as $T_n \to +\infty$, to the process $\widehat{\varphi}\left(\widetilde{\zeta}(t)\right)$, where $\widehat{\varphi}(x)$ is the inverse function to $\widehat{f}(x)$. The process $\xi_{T_n}(t)$ is weakly equivalent to the process $\widetilde{\xi}_{T_n}(t)$ for every T_n, and, according to the conditions of Theorem 2.3, we have that $\widehat{\varphi}\left(\widetilde{\zeta}(t)\right) = l\left(\zeta(t)\right)$, where $\zeta(t)$ is the solution of the equation

$$\zeta(t) = \int_0^t \overline{\sigma}\left(\zeta(s)\right) d\widehat{W}(s).$$

Note that $\zeta(t) = l^{-1}\left(\widehat{\varphi}\left(\widetilde{\zeta}(t)\right)\right)$ and $\overline{\sigma}\left(\zeta(t)\right) = \overline{\sigma}\left(\widetilde{\zeta}(t)\right)$. So, taking into account equality (2.29), we have

$$l^{-1}\left(\widehat{\varphi}\left(\widetilde{\zeta}(t)\right)\right) = \int_0^t \overline{\sigma}\left(\widetilde{\zeta}(s)\right)\sqrt{Q'\left(\widetilde{\zeta}(s)\right)}\, d\widetilde{\zeta}(s)$$

with probability 1 for all $t > 0$. Therefore, according to Lemma A.12, we obtain

$$\frac{l^{-1}\left(\widehat{\varphi}(x)\right)}{x} = \overline{\sigma}(x)\sqrt{Q'(x)} = k_0$$

for all x, where k_0 is some constant. Thus,

$$Q'(x) = \left(\frac{k_0}{\overline{\sigma}(x)}\right)^2, \quad \widehat{f}(x) = x\frac{\overline{c}(x)}{k_0}.$$

In addition, the following convergence holds:

$$f_{T_n}(x) = \frac{f\left(x\sqrt{T_n}\right)}{\sqrt{T_n}} = x\frac{f\left(x\sqrt{T_n}\right)}{x\sqrt{T_n}} \to x\frac{\overline{c}(x)}{k_0},$$

and

$$Q_{T_n}(x) = \int_0^x \frac{dv}{\left[f'\left(\varphi\left(v\sqrt{T_n}\right)\right)\sigma\left(\varphi\left(v\sqrt{T_n}\right)\right)\right]^2}$$

$$= \frac{1}{\sqrt{T_n}}\int_0^{\varphi(x\sqrt{T_n})} \frac{dv}{f'(v)\sigma^2(v)}$$

$$= x\frac{1}{f\left(\varphi\left(x\sqrt{T_n}\right)\right)}\int_0^{\varphi(x\sqrt{T_n})} \frac{dv}{f'(v)\sigma^2(v)} \to x\frac{k_0^2}{\sigma^2(x)},$$

as $T_n \to +\infty$.

Since the subsequence $T_n \to +\infty$ is arbitrary, the proof of the necessity with $k = \frac{1}{k_0}$ is complete.   □

Consider an important partial case of Theorem 2.3.

Theorem 2.4 *Let ξ be a solution of Eq. (2.1), and let*

$$0 < \delta \le \sigma(x) \le C, \quad \left|\int_0^x \frac{a(u)}{\sigma^2(u)}du\right| \le C.$$

Then the stochastic process $\xi_T(t) = \frac{\xi(tT)}{\sqrt{T}}$ converges weakly, as $T \to +\infty$, to the process $\sigma_0 W(t)$, where W is a Wiener process, if and only if

$$\frac{f(x)}{x} \to \widehat{\sigma}_1, \quad \frac{1}{x}\int_0^x \frac{du}{f'(u)\sigma^2(u)} \to \widehat{\sigma}_2,$$

as $|x| \to +\infty$, with $\sigma_0 = (\widehat{\sigma}_1\widehat{\sigma}_2)^{-\frac{1}{2}}$.

It is easy to see that the latter assertion follows from Theorem 2.3 with $c_1 = c_2$ and $\frac{\sigma_1}{c_1} = \frac{\sigma_2}{c_2} = \sigma_0$ (or from Theorem 2.2).

We note that sufficient conditions of Theorem 2.4 are obtained in [31], and necessary conditions of Theorem 2.4 derive from the paper [41].

Remark 2.7 It is clear that the process $\xi_T(t) = \frac{\xi(tT)}{B(T)}$, which is considered in Lemma 2.3, is a solution of the equation

$$d\xi_T(t) = a_T(\xi_T(t))\,dt + \sigma_T(\xi_T(t))\,dW_T(t),$$

where

$$a_T(x) = \frac{T}{B(T)}a(xB(T)), \quad \sigma_T(x) = \frac{\sqrt{T}}{B(T)}\sigma(xB(T)).$$

In particular, the following cases are possible:

(1) $\dfrac{\sqrt{T}}{B(T)} \to +\infty;$ (2) $\dfrac{\sqrt{T}}{B(T)} \to 0;$ (3) $B(T) \sim \sqrt{T},$ as $T \to +\infty.$

Therefore, $a_T(x)$ and $\sigma_T^2(x)$ can be sequences of "δ"-type as well as can have degeneration of another nature (see Example 2.6).

2.4 Examples

Consider the following examples of the coefficients $a(x)$ and $\sigma(x)$ in Eq. (2.1).

Example 2.1 Let $a(x) = xe^{-x^2}$ and $\sigma(x) = 1$.

Since $\int\limits_{\mathbb{R}} xe^{-x^2}dx = 0$, then $\int\limits_{0}^{+\infty} xe^{-x^2}dx = \int\limits_{0}^{-\infty} xe^{-x^2}dx$ and

$$f'(x) = \exp\left\{-2\int\limits_0^x a(v)\,dv\right\} \to \hat{\sigma}_1 = \exp\left\{-2\int\limits_0^{+\infty} a(v)\,dv\right\},$$

as $|x| \to +\infty$.
Therefore

$$\frac{f(x)}{x} \to \hat{\sigma}_1, \quad \frac{1}{x}\int\limits_0^x \frac{dv}{f'(v)} \to \frac{1}{\hat{\sigma}_1},$$

as $|x| \to +\infty$.

Here $\widehat{\sigma}_2 = \widehat{\sigma}_1^{-1}$ and $\sigma_0 = (\widehat{\sigma}_1\widehat{\sigma}_2)^{-\frac{1}{2}} = 1$.

Consequently, according to Theorem 2.4, the stochastic process $\xi_T(t) = \xi(tT)T^{-\frac{1}{2}}$ converges weakly, as $T \to +\infty$, to the Wiener process W.

Example 2.2 Let $a(x) = \sin \beta x$, $\sigma(x) = 1$, $\beta \neq 0$.

Then

$$f'(x) = \exp\left\{-2\int_0^x a(v)\,dv\right\} = e^{-\frac{2}{\beta}}e^{\frac{2}{\beta}\cos \beta x}.$$

Therefore

$$\frac{1}{x}\int_0^x f'(u)\,du = \frac{1}{x}\int_0^x e^{-2\int_0^u a(v)dv}\,du = e^{-\frac{2}{\beta}}\frac{1}{x}\int_0^x e^{\frac{2}{\beta}\cos \beta u}\,du \to e^{-\frac{2}{\beta}}\frac{1}{2\pi}\int_0^{2\pi} e^{\frac{2}{\beta}\cos z}\,dz = \widehat{\sigma}_1,$$

as $|x| \to +\infty$, and

$$\frac{1}{x}\int_0^x \frac{du}{f'(u)} = \frac{1}{x}\int_0^x e^{2\int_0^u a(v)dv}\,du \to e^{\frac{2}{\beta}}\frac{1}{2\pi}\int_0^{2\pi} e^{-\frac{2}{\beta}\cos z}\,dz = \widehat{\sigma}_2,$$

as $|x| \to +\infty$.

Taking into account the equalities

$$\int_0^{2\pi} e^{\frac{2}{\beta}\cos z}\,dz = \int_0^{2\pi} e^{-\frac{2}{\beta}\cos z}\,dz \quad \text{and} \quad \int_0^{\pi} e^{\frac{2}{\beta}\cos z}\,dz = \int_{\pi}^{2\pi} e^{\frac{2}{\beta}\cos z}\,dz,$$

we get that

$$\sigma_0 = (\widehat{\sigma}_1\widehat{\sigma}_2)^{-\frac{1}{2}} = \left[\frac{1}{2\pi}\int_0^{2\pi} e^{\frac{2}{\beta}\cos z}\,dz\right]^{-1} = \left[\frac{1}{\pi}\int_0^{\pi} e^{\frac{2}{\beta}\cos z}\,dz\right]^{-1}.$$

According to Theorem 2.4, the stochastic process $\xi(tT)T^{-\frac{1}{2}}$ converges weakly, as $T \to +\infty$, to the process $\sigma_0 W(t)$.

Note that the period of the periodic drift coefficient asymptotically affects the diffusion coefficient: if the period of oscillation decreases, the coefficient of diffusion increases asymptotically, while the drift coefficient is averaged and equals zero in the equation for the limit process.

Example 2.3 Let $a(x) = 0$, $\sigma(x) = \frac{1}{2+\cos \beta x}$.

In this case

$$\frac{1}{x}\int_0^x e^{-2\int_0^u \frac{a(v)}{\sigma^2(v)}dv}\,du = 1, \quad \text{for } x \neq 0,$$

and

$$\frac{1}{x}\int_0^x e^{2\int_0^u \frac{a(v)}{\sigma^2(v)}dv}\frac{1}{\sigma^2(u)}\,du = \frac{1}{x}\int_0^x [2+\cos\beta u]^2\,du$$

$$= 4 + \frac{4}{x}\int_0^x \cos\beta u\,du + \frac{1}{x}\int_0^x \cos^2\beta u\,du$$

$$= 4 + \frac{1}{2}\frac{1}{x}\int_0^x [1+\cos 2\beta u]\,du + o(1) = 4 + \frac{1}{2} + o(1) \to \frac{9}{2},$$

as $|x| \to +\infty$. Therefore, according to Theorem 2.4 with $\sigma_0 = \frac{\sqrt{2}}{3}$, the stochastic process $\xi(tT)T^{-\frac{1}{2}}$ converges weakly, as $T \to +\infty$, to the process $\frac{\sqrt{2}}{3}W(t)$. Consequently, the magnitude of the period of the periodic diffusion coefficient does not asymptotically affect the diffusion coefficient in the equation of the limit process.

Example 2.4 $a(x) = xe^{-x^2}$, $\sigma(x) = \frac{1}{2+\cos\beta x}$.
 In this case

$$\frac{1}{x}\int_0^x e^{-2\int_0^u \frac{a(v)}{\sigma^2(v)}dv}\,du \to c_0 = \exp\left\{-2\int_0^{+\infty}\frac{a(v)}{\sigma^2(v)}dv\right\},$$

as $|x| \to +\infty$, and, using the previous example, we have

$$\frac{1}{x}\int_0^x e^{2\int_0^u \frac{a(v)}{\sigma^2(v)}dv}\frac{1}{\sigma^2(u)}\,du = \frac{1}{c_0}\frac{1}{x}\int_0^x [2+\cos\beta u]^2\,du + o(1) \to \frac{9}{2c_0},$$

as $|x| \to +\infty$. According to Remark 2.6 with $c_1 = c_2 = c_0$ and $\frac{\sigma_1}{c_1} = \frac{\sigma_2}{c_2} = \frac{\sqrt{2}}{3}$, the stochastic process $\xi(tT)T^{-\frac{1}{2}}$ converges weakly, as $T \to +\infty$, to the process $\frac{\sqrt{2}}{3}W(t)$.

Example 2.5

(1) Let $a(x) = \frac{1}{1+x^2}$, $\sigma(x) = 1$.

Since

$$\int\limits_{0}^{+\infty} a(x)dx = \frac{\pi}{2}, \quad \int\limits_{0}^{-\infty} a(x)dx = -\frac{\pi}{2},$$

then, according to Theorem 2.3, the stochastic process $\xi_T(t) = \xi(tT)T^{-\frac{1}{2}}$ converges weakly, as $T \to +\infty$, to the process $\widehat{\xi}(t) = \ell(\zeta(t))$, where $\zeta(t)$ is a solution to the Itô equation $d\zeta(t) = \overline{\sigma}(\zeta(t))\, dW(t)$, where

$$\overline{\sigma}(x) = \begin{cases} e^{-\pi}, & x \geq 0, \\ e^{\pi}, & x < 0, \end{cases} \qquad \ell(x) = \frac{x}{\overline{\sigma}(x)}.$$

Using the explicit form (2.10) of the transition density $\rho_\zeta(t, x, y)$ of the process $\zeta(t)$, we obtain an explicit form of the transition density of the process $\widehat{\xi}(t)$:

$$\rho_{\widehat{\xi}}(t, x, y) = \rho_\zeta(t, \ell^{-1}(x), \ell^{-1}(y))(\ell^{-1}(y))'$$

$$= \begin{cases} \dfrac{1}{\sqrt{2\pi t}}\left[e^{-\frac{(y-x)^2}{2t}} + \dfrac{\sigma_2 - \sigma_1}{\sigma_2 + \sigma_1} \cdot e^{-\frac{(y+x)^2}{2t}} \right], & x \geq 0, \ y > 0; \\[3mm] \dfrac{2\sigma_1}{\sigma_1 + \sigma_2}\dfrac{1}{\sqrt{2\pi t}} e^{-\frac{(y-x)^2}{2t}}, & x \geq 0, \ y < 0; \\[3mm] \dfrac{2\sigma_2}{\sigma_1 + \sigma_2}\dfrac{1}{\sqrt{2\pi t}} e^{-\frac{(y-x)^2}{2t}}, & x \leq 0, \ y > 0; \\[3mm] \dfrac{1}{\sqrt{2\pi t}}\left[e^{-\frac{(y-x)^2}{2t}} - \dfrac{\sigma_2 - \sigma_1}{\sigma_2 + \sigma_1} \cdot e^{-\frac{(y+x)^2}{2t}} \right], & x \leq 0, \ y < 0, \end{cases}$$

where $\sigma_1 = e^{-\pi}$, $\sigma_2 = e^{\pi}$.

It is easy to get that for the limit process $\widehat{\xi}(t)$ there exist Kolmogorov's local characteristics in the generalized sense: the diffusion coefficient $\sigma(x) = 1$ and the drift coefficient is $c\delta(\cdot)$, where $\delta(\cdot)$ is Dirac's delta function, $c = \frac{\sigma_2 - \sigma_1}{\sigma_2 + \sigma_1} = \tanh \pi$. Furthermore, we can formally write (see [42]) the Itô SDE for the process $\widehat{\xi}(t)$ as

$$d\widehat{\xi}(t) = \tanh\pi\, \delta(\cdot)\, dt + dW(t).$$

Note that in this case the diffusion coefficient in the differential equation for the process $\xi_T(t)$ equals to 1, and the drift coefficient $a_T(x) = \sqrt{T}a(x\sqrt{T})$ is a "δ"-type sequence. After passing to the limit formally in the equation for the process $\xi_T(t)$ we obtain the drift coefficient $\pi\delta(\cdot)$ instead of $\tanh \pi\delta(\cdot)$ for the limit process $\widehat{\xi}(t)$. This fact emphasizes the specificity of passing to the limit in the Itô SDEs. See Appendix A.3 for more details.

(2) Let

$$a(x) = \frac{\bar{a}(x)}{1 + x^2}, \quad \bar{a}(x) = \begin{cases} a_1, & x \geq 0, \\ a_2, & x < 0, \end{cases} \quad \sigma(x) = 1.$$

In this case all the conclusions from (1) hold with $\sigma_1 = e^{-a_1\pi}$ for $x \geq 0$, $\sigma_2 = e^{-a_2\pi}$ for $x < 0$.

The limit process $\widehat{\xi}(t)$ satisfies the Itô SDE

$$d\widehat{\xi}(t) = c_0 \, \delta(\cdot)dt + dW(t),$$

where $c_0 = \frac{\sigma_2 - \sigma_1}{\sigma_2 + \sigma_1}$.

In particular, in the case where

(2.1) $a_2 = -a_1$ we have $c_0 = 0$ and the stochastic process $\xi_T(t) = \xi(tT)T^{-\frac{1}{2}}$ converges weakly, as $T \to +\infty$, to the Wiener process W. We have here $\int_{\mathbb{R}} a(x)dx = 0$;

(2.2) $a_2 = a_1 = a_0$ we have $c_0 = \tanh a_0\pi$ and the stochastic process $\xi_T(t)$ converges weakly, as $T \to +\infty$, to the solution of the equation

$$d\widehat{\xi}(t) = \tanh a_0\pi \, \delta(\cdot)dt + dW(t).$$

We have here $\int_{\mathbb{R}} a(x)dx = a_0\pi$.

Example 2.6

(1) Let $a(x) = -\frac{x}{(1+x^2)^3}$, $\sigma(x) = \frac{1}{1+x^2}$.

Note that

$$\int_0^x \frac{a(v)}{\sigma^2(v)} \, dv = -\int_0^x \frac{v}{1 + v^2} \, dv = -\frac{1}{2} \ln(1 + x^2)$$

and

$$f'(x) = \exp\left\{-2\int_0^x \frac{a(v)}{\sigma^2(v)} \, dv\right\} = 1 + x^2,$$

then $f'(x)\sigma(x) = 1$. The conditions of Theorem 2.1 are fulfilled with $\sigma_1 = \sigma_2 = 1$ and the stochastic process $f(\xi(tT))T^{-\frac{1}{2}}$ converges weakly, as $T \to +\infty$, to the Wiener process $W(t)$. Moreover,

$$f(x) = \int_0^x f'(x)\,dx = x + \frac{x^3}{3} \quad \text{and} \quad \lim_{k \to +\infty} \frac{f(kx)}{f(k)} = x^3.$$

According to Lemma 2.3, the finite-dimensional distributions of the process $\xi_T(t) = \xi(tT)B^{-1}(T)$, where $B(T) = c_0 T^{\frac{1}{6}}$, $c_0 = \sqrt[3]{3}$, converge, as $T \to +\infty$, to the corresponding finite-dimensional distributions of the process $\sqrt[3]{W(t)}$ with the transition density

$$p(t, x, y) = \frac{3y^2}{\sqrt{2\pi t}} e^{-\frac{(y^3 - x^3)^2}{2t}}.$$

Note that the process $\xi_T(t)$ for every fixed $T > 0$ satisfies the following Itô SDE (see Remark 2.7)

$$d\xi_T(t) = a_T\left(\xi_T(t)\right)dt + \sigma_T\left(\xi_T(t)\right)dW_T(t), \quad \xi_T(0) = 0, \tag{2.30}$$

where $W_T(t) = \frac{W(tT)}{\sqrt{T}}$ is a Wiener process for every fixed $T > 0$,

$$a_T(x) = \frac{T}{B(T)}a(xB(T)) = -\frac{T}{B(T)}\frac{xB(T)}{\left(1 + x^2 B^2(T)\right)^3} = -\frac{Tx}{\left(1 + x^2 c_0^2 T^{\frac{1}{3}}\right)^3}$$

$$= -\frac{T^{\frac{5}{6}}x}{c_0} \cdot \widehat{a}_T(x) \quad \text{with} \quad \widehat{a}_T(x) = \frac{c_0 T^{\frac{1}{6}}}{\left(1 + x^2 c_0^2 T^{\frac{1}{3}}\right)^3}.$$

Note that for any continuous function $\varphi(x)$ with compact support

$$\int_{\mathbb{R}} \varphi(x)\widehat{a}_T(x)\,dx = \int_{\mathbb{R}} \varphi\left(\frac{z}{c_0 T^{\frac{1}{6}}}\right) \frac{dz}{(1 + z^2)^3} \to \varphi(0)m_0,$$

as $T \to +\infty$, where $m_0 = \int_{\mathbb{R}} \frac{dz}{(1 + z^2)^3}$ is the weight of the "δ"-type sequence $\widehat{a}_T(x)$ at the point $x = 0$.

The diffusion coefficient in Eq. (2.30) has the form

$$\sigma_T(x) = \frac{\sqrt{T}}{B(T)}\sigma(xB(T)) = \frac{\sqrt{T}}{c_0 T^{\frac{1}{6}}}\frac{1}{1 + x^2 c_0^2 T^{\frac{1}{3}}} = \frac{T^{\frac{1}{6}}}{c_0^2} \cdot \widehat{\sigma}_T(x)$$

with

$$\widehat{\sigma}_T(x) = \frac{c_0 T^{\frac{1}{6}}}{1 + x^2 c_0^2 T^{\frac{1}{3}}}.$$

It is a "δ"-type sequence at the point $x = 0$ with weight $m_0 = \int_{\mathbb{R}} \frac{dz}{1+z^2} = \pi$.

In this case

$$f_T(x) = \int_0^x \exp\left\{-2\int_0^u \frac{a_T(v)}{\sigma_T^2(v)} \, dv\right\} du = \frac{c_0 x}{T^{\frac{1}{3}}} + x^3.$$

(2) Let the diffusion coefficient in Eq. (2.30) equal

$$\sigma_T(x) = \frac{T^{\frac{1}{6}}}{1 + x^2 T^{\frac{1}{3}}}.$$

It is a "δ"-type sequence at the point $x = 0$ with weight π. The drift coefficient $a_T(x)$ will be found from the equality $f_T'(x)\sigma_T(x) = 1$, where $f_T'(x) = \exp\left\{-2\int_0^x \frac{a_T(v)}{\sigma_T^2(v)} \, dv\right\}$.

As a result of differentiation of the equality

$$-2\int_0^x \frac{a_T(v)}{\sigma_T^2(v)} \, dv = \ln \frac{1}{\sigma_T(x)}$$

we get that

$$a_T(x) = \frac{1}{2}\sigma_T(x)\sigma_T'(x).$$

Thus,

$$a_T(x) = \frac{1}{2}\frac{T^{\frac{1}{6}}}{1 + x^2 T^{\frac{1}{3}}} \cdot \left(-T^{\frac{1}{6}}\frac{2x T^{\frac{1}{3}}}{\left(1 + x^2 T^{\frac{1}{3}}\right)^2}\right)$$

$$= -\frac{x T^{\frac{2}{3}}}{\left(1 + x^2 T^{\frac{1}{3}}\right)^3} = -x T^{\frac{1}{2}}\widehat{a}_T(x) \quad \text{with } \widehat{a}_T(x) = \frac{T^{\frac{1}{6}}}{\left(1 + x^2 T^{\frac{1}{3}}\right)^2},$$

which is a "δ"-type sequence at the point $x = 0$ with weight $\int_{\mathbb{R}} \frac{dz}{(1+z^2)^2}$.

In this case $\zeta_T(t) = f_T(\xi_T(t)) = W_T(t)$ converges weakly, as $T \to +\infty$, to the Wiener process $W(t)$. Here $f_T(x) = \frac{1}{3}x^3 T^{\frac{1}{3}} + o(1)$, where $o(1) \to 0$, as $T \to +\infty$, for all $x \in \mathbb{R}$.

(3) Let the diffusion coefficient in Eq. (2.30) equal

$$\sigma_T(x) = \sqrt{\frac{T^{\frac{1}{6}}}{1 + x^2 T^{\frac{1}{3}}}}.$$

Note that $\sigma_T^2(x)$ is a "δ"-type sequence at the point $x = 0$ with weight π. The drift coefficient $a_T(x)$ will be found from the equality $f_T'(x)\sigma_T(x) = 1$ similarly to (2).

Thus,

$$a_T(x) = \frac{1}{2}\sigma_T(x)\sigma_T'(x) = \frac{1}{2}\frac{T^{\frac{1}{12}}}{\left(1 + x^2 T^{\frac{1}{3}}\right)^{\frac{1}{2}}} \cdot \left(-\frac{2x T^{\frac{5}{12}}}{\left(1 + x^2 T^{\frac{1}{3}}\right)^{\frac{3}{2}}}\right)$$

$$= -\frac{x T^{\frac{1}{2}}}{\left(1 + x^2 T^{\frac{1}{3}}\right)^2} = -x T^{\frac{1}{3}} \widehat{a}_T(x) \quad \text{with } \widehat{a}_T(x) = \frac{T^{\frac{1}{6}}}{\left(1 + x^2 T^{\frac{1}{3}}\right)^2},$$

which is a "δ"-type sequence at the point $x = 0$ with weight $m_0 = \int_{\mathbb{R}} \frac{dz}{(1+z^2)^2}$.

Consequently, $\zeta_T(t) = W_T(t)$ converges weakly, as $T \to +\infty$, to the Wiener process W. In this case $f_T(x) = \frac{x^2}{2} T^{\frac{1}{12}} \operatorname{sign} x + o(1)$.

Chapter 3
Asymptotic Analysis of Equations with Ergodic and Stochastically Unstable Solutions

In this chapter, we consider one-dimensional homogeneous stochastic differential equations whose coefficients place these equations on the border between equations whose solutions have ergodic distribution, and equations with stochastically unstable solutions. To simplify calculations and to visualize better the influence of the drift coefficient of the equation on the asymptotic behavior of solution, we consider Eq. (2.1) with $\sigma(x) \equiv 1$. Statements about the instability and ergodicity for the solutions are formulated and proved in Sect. 3.1. Weak convergence of normalized stochastically unstable solutions to the Bessel diffusion process we consider in Sect. 3.2. Section 3.3 includes more general results about the influence of the coefficients of the equation on the limit behavior of the solutions. Influence of the diffusion coefficient on the limit behavior of the stochastically unstable solutions we study in Sect. 3.4. Section 3.5 contains several examples.

3.1 Criteria of Instability and Ergodicity for the Solutions

Let us consider Eq. (2.1) with $\sigma(x) \equiv 1$, namely an equation of the form

$$d\xi(t) = a(\xi(t))dt + dW(t), \quad t > 0, \quad \xi(0) = x_0, \tag{3.1}$$

with real measurable drift coefficient satisfying additional assumption: $|x\,a(x)| \leq L$ for a certain constant L and for all $x \in \mathbb{R}$. In particular, it can be

$$a(x) \sim \frac{c}{|x|},$$

© Springer Nature Switzerland AG 2020
G. Kulinich et al., *Asymptotic Analysis of Unstable Solutions of Stochastic Differential Equations*, Bocconi & Springer Series 9,
https://doi.org/10.1007/978-3-030-41291-3_3

as $|x| \to +\infty$. The behavior of the solutions ξ to the class of equations of type (3.1), in which

$$a\,(x) \sim \frac{c}{|x|^\alpha},$$

as $|x| \to +\infty$, for $\alpha > 1$, was investigated in Chap. 2. Indeed, in this case $\int_{\mathbb{R}} |a(u)|\, du < +\infty$ and for all $x \in \mathbb{R}$ we obtain that $\left| \int_0^x a(u)\, du \right| \le C$. Consequently, the conditions (2.4) and (2.11) hold.

For $-1 < \alpha < 1$ the solutions of Eq. (3.1) have an exact order of behavior, as $t \to +\infty$, i.e., $\xi(t) \to +\infty$ with probability 1, as $t \to +\infty$, and there exists a non-random function $B(t) \to +\infty$, as $t \to +\infty$, such that

$$\mathsf{P}\left\{ \lim_{t \to +\infty} \frac{\xi\,(t)}{B\,(t)} = 1 \right\} = 1.$$

For more details see the book [17, § 17].

It is well known (see [82, Theorem 4]) that the SDE (3.1) possesses a unique strong pathwise solution and this solution is a homogeneous strong Markov process.

In this chapter we use the following notations:

$$f(x) = \int_0^x \exp\left\{ -2 \int_0^u a(v)dv \right\} du$$

and

$$\psi(x, c) = \frac{1}{\ln |x|} \int_0^x a(v)dv - c. \tag{3.2}$$

In what follows the constants $C > 0$ and $L > 0$ do not depend on x and t.

Theorem 3.1 *Let ξ be a solution to Eq. (3.1) and let*

$$\lim_{|x| \to +\infty} \psi(x, c_0) = 0. \tag{3.3}$$

Then we have the following cases, depending on the value of c_0.

1. For $2c_0 < -1$, the solution ξ is ergodic and

$$\mathsf{P}\{\xi(t) < x\} \to \left[\int_{\mathbb{R}} \frac{dv}{f'(v)} \right]^{-1} \int_{-\infty}^x \frac{dv}{f'(v)},$$

as $t \to +\infty$.

2. *For $2c_0 > -1$, the solution ξ is stochastically unstable, in other words*

$$\lim_{t\to+\infty} \frac{1}{t} \int_0^t \mathbf{P}\{|\xi(s)| < N\}\, ds = 0$$

for any constant $N > 0$.

The case $c_0 = -\frac{1}{2}$ will be considered in Theorem 3.2.

Proof

Statement 1 For $x \neq 0$ we have the representation

$$f'(x) = |x|^{-2c_0} \exp\{-2\ln|x|\,\psi(x, c_0)\}. \tag{3.4}$$

In the case $2c_0 < -1$, we can choose $\varepsilon > 0$ such that $2c_0 + 2\varepsilon < -1$. According to relation (3.3), there exists a constant $L > 0$ such that $|\psi(x, c_0)| < \varepsilon$ for $|x| > L$. Using representation (3.4), we obtain the inequality $f'(x) \geq |x|^{-2c_0}|x|^{-2\varepsilon}$ for $|x| > L$.

Therefore, we have that $f(-\infty) = -\infty$, $f(+\infty) = +\infty$ and

$$\int_{\mathbb{R}} \frac{dx}{f'(x)} < +\infty. \tag{3.5}$$

Consider the process $\eta(t) = f(\xi(t))$. Using the Itô formula, we conclude that

$$d\eta(t) = \hat{\sigma}(\eta(t))dW(t),$$

where $\hat{\sigma}(x) = f'(\varphi(x))$ and the function $\varphi(x)$ is the inverse function to $f(x)$. It is clear that the function $\varphi(x)$ is monotonically increasing and $\varphi(-\infty) = -\infty$, $\varphi(+\infty) = +\infty$.

Using the substitution rule for the integral

$$\int_{\mathbb{R}} \frac{dx}{\hat{\sigma}^2(x)} = \int_{\mathbb{R}} \frac{dx}{[f'(\varphi(x))]^2}$$

with $u = \varphi(x)$, $du = \frac{dx}{f'(x)}$ we obtain the equality

$$\int_{\mathbb{R}} \frac{dx}{\hat{\sigma}^2(x)} = \int_{\mathbb{R}} \frac{du}{f'(u)}.$$

Taking into account the boundedness of the function $a(x)$, the last equality and (3.5) we have (see Theorem A.10) that the process $\eta(t)$ is ergodic and

$$P\{\eta(t) < x\} \to \left[\int_{\mathbb{R}} \frac{dv}{f'(v)}\right]^{-1} \int_{-\infty}^{\varphi(x)} \frac{dv}{f'(v)},$$

as $t \to +\infty$. Statement 1 follows from the last relation.

Statement 2 Consider the function

$$\Phi(x) = 2 \int_0^x \left\{ f'(u) \int_0^u \left(\frac{\chi_{|v|<N}}{f'(v)}\right) dv \right\} du.$$

Note that

$$\frac{1}{2}\Phi''(x) + \Phi'(x)a(x) = \chi_{|x|<N}$$

a.e. with respect to the Lebesgue measure. Therefore, using the Itô formula (Lemma A.3), we conclude that

$$\frac{1}{t} \int_0^t P\{|\xi(s)| < N\}\, ds = \frac{1}{t} E\left[\Phi(\xi(t)) - \Phi(\xi(0))\right]. \tag{3.6}$$

Let us prove that for $2c_0 > -1$ we have the convergence

$$\lim_{|x| \to +\infty} \frac{\Phi(x)}{x^2} = 0. \tag{3.7}$$

In order to do this, consider $\varepsilon > 0$ such that $-2c_0 + 2\varepsilon < 1$. Using representation (3.4), we obtain the inequality $f'(x) \leq |x|^{-2c_0+2\varepsilon}$ for $|x| > L$.
Therefore,

$$\frac{f'(x)}{x} \to 0 \quad \text{and} \quad \frac{\Phi(x)}{x^2} \to 0,$$

as $|x| \to +\infty$.

So, for arbitrary $\varepsilon > 0$ there exists a constant L_ε such that $\left|x^{-2}\Phi(x)\right| < \varepsilon$ for $|x| > L_\varepsilon$. Consequently,

$$\frac{1}{t} E\left|\Phi(\xi(t))\right| \leq \frac{C_\varepsilon}{t} + \varepsilon \frac{1}{t} E\xi^2(t),$$

where $C_\varepsilon = \sup\limits_{|x|<L_\varepsilon} |\Phi(x)|$. Furthermore,

$$\mathsf{E}\xi^2(t) = x_0^2 + \mathsf{E}\int_0^t [2\xi(s)a(\xi(s)) + 1]\,ds \le x_0^2 + Ct.$$

Thus,

$$\limsup_{t\to+\infty} \frac{1}{t}\mathsf{E}\,|\Phi(\xi(t))| \le C\cdot\varepsilon.$$

Hence, we have

$$\lim_{t\to+\infty} \frac{1}{t}\mathsf{E}\,|\Phi(\xi(t))| = 0. \tag{3.8}$$

Taking into account (3.6) and (3.8), we obtain the proof of statement 2.

$\square$

Consider now the case $2c_0 = -1$, applying some of previous calculations.

Theorem 3.2 *Let ξ be a solution to Eq. (3.1) and let (3.3) hold for $2c_0 = -1$. Then we have*

1. *If $\int_{\mathbb{R}} \frac{dx}{f'(x)} < +\infty$, then the solution ξ is ergodic and*

$$\mathsf{P}\{\xi(t) < x\} \to \left[\int_{\mathbb{R}} \frac{dv}{f'(v)}\right]^{-1} \int_{-\infty}^x \frac{dv}{f'(v)},$$

   as $t \to +\infty$.
2. *If $x^{-1}f'(x) \to 0$, as $|x| \to +\infty$, then the solution ξ is stochastically unstable, in other words*

$$\lim_{t\to+\infty} \frac{1}{t}\int_0^t \mathsf{P}\{|\xi(s)| < N\}\,ds = 0$$

   for any constant $N > 0$.

Proof Since the inequality $2c_0 < -1$ is used only in the proof of relation (3.5) in Theorem 3.1, the further considerations in the proof of statement 1 in Theorem 3.1 can be fully used also here.

The inequality $2c_0 > -1$ is used only in the proof of the convergence $x^{-1}f'(x) \to 0$, as $|x| \to +\infty$ in Theorem 3.1. The further considerations in the proof of statement 2 in Theorem 3.1 can be fully used also here. $\square$

Remark 3.1 It follows from Theorem 3.2 that the case $2c_0 = -1$ is critical. Actually, for $2c_0 = -1$ we have equations with ergodic solutions and equations with stochastically unstable solutions depending on the rate of convergence (3.3).

3.2   Convergence of Normalized Stochastically Unstable Solutions to the Bessel Diffusion Process

Next, consider the case $2c_0 > -1$ and investigate the asymptotic behavior of stochastically unstable solutions.

Theorem 3.3 *Let ξ be a solution to Eq. (3.1), and let there exist the constants c_1 and c_2 such that*

$$\lim_{|x| \to +\infty} \left[\frac{1}{x} \int_0^x va(v)dv - \bar{c}(x) \right] = 0, \quad \bar{c}(x) = \begin{cases} c_1, & x \geq 0, \\ c_2, & x < 0. \end{cases} \quad (3.9)$$

Then we have three cases.

(1) *If $c_1 = c_2 = c_0$, $2c_0 > -1$, then the stochastic process $|\xi(tT)| T^{-\frac{1}{2}}$ converges weakly, as $T \to +\infty$, to the process $r(t)$, which is the solution of Itô's SDE*

$$r^2(t) = (2c+1)t + 2 \int_0^t r(s)dW(s) \quad (3.10)$$

for $c = c_0$;

(2) *if $2c_1 > 1$ and $2c_2 < 1$, then the stochastic process $\xi(tT)T^{-\frac{1}{2}}$ converges weakly, as $T \to +\infty$, to the solution $r(t)$ to Eq. (3.10) for $c = c_1$;*

(3) *if $2c_1 < 1$ and $2c_2 > 1$, then the stochastic process $-\xi(tT)T^{-\frac{1}{2}}$ converges weakly, as $T \to +\infty$, to the solution $r(t)$ to Eq. (3.10) for $c = c_2$.*

Definition 3.1 A nonnegative homogeneous Markov process ζ with transition density of the form

$$p(t, x, y) = \frac{1}{t \, (xy)^{\nu-1}} \exp \left\{ -\frac{x^2 + y^2}{2t} \right\} y^{2\nu-1} I_{\nu-1} \left(\frac{xy}{t} \right),$$

where I_ν is the modified Bessel function, is called a Bessel diffusion process of index $\nu > 0$.

This process ζ is a solution of Itô's SDE

$$\zeta^2(t) = \zeta^2(0) + \nu t + 2 \int_0^t \zeta(s)dW(s).$$

Remark 3.2 The process $r(t)$, which is the solution to Eq. (3.10), is the Bessel diffusion process of index $\nu = 2c + 1$ (see Definition 3.1). Note that for $-1 < 2c < 1$, $r(t)$ is a process with reflection at the origin, and for $2c \geq 1$ the origin is not attainable by the process $r(t)$. See Example IV–8.3 in [23] and [24, 77] for further information on Bessel diffusions.

To prove Theorem 3.3 we need the following statement.

Lemma 3.1 *If condition (3.9) holds, then*

$$\lim_{|x| \to +\infty} \psi(x, \bar{c}(x)) = 0. \tag{3.11}$$

Proof Let us first consider the limit for $x \to +\infty$. For arbitrary $\varepsilon > 0$ there exists a constant $C_\varepsilon > 0$ such that, for $x > C_\varepsilon$, we have the inequality

$$\left| \frac{1}{x} \int_{C_\varepsilon}^{x} v a(v) \, dv - c_1 \right| < \varepsilon.$$

Then for $x > C_\varepsilon$

$$I(x) := \frac{1}{\ln x} \int_0^x a(v) \, dv - c_1 = \frac{1}{\ln x} \int_0^{C_\varepsilon} a(v) \, dv + \frac{1}{\ln x} \int_{C_\varepsilon}^x a(v) \, dv - c_1$$

$$= o(1) + \frac{1}{\ln x} c_1 \int_{C_\varepsilon}^x \frac{1}{v} \, dv + \frac{1}{\ln x} \int_{C_\varepsilon}^x \frac{v a(v) - c_1}{v} \, dv - c_1$$

$$= o(1) + \frac{1}{\ln x} \int_{C_\varepsilon}^x \frac{v a(v) - c_1}{v} \, dv$$

$$= o(1) + \frac{1}{\ln x} \left[\frac{1}{x} \int_{C_\varepsilon}^x (z a(z) - c_1) \, dz + \int_{C_\varepsilon}^x \frac{1}{v^2} \left(\int_{C_\varepsilon}^v (z a(z) - c_1) \, dz \right) dv \right]$$

$$= o(1) + \frac{1}{\ln x} \left[\frac{1}{x} \int_{C_\varepsilon}^x (z a(z) - c_1) \, dz \right] + \frac{1}{\ln x} \int_{C_\varepsilon}^x \frac{1}{v^2} \int_{C_\varepsilon}^v (z a(z) - c_1) \, dz \, dv.$$

So, for $x > C_\varepsilon$,

$$|I(x)| \le o(1) + \frac{1}{\ln x} [\varepsilon (\ln x - \ln C_\varepsilon)] = o(1) + \varepsilon.$$

Consequently,

$$\limsup_{x \to +\infty} |I(x)| < \varepsilon$$

for any $\varepsilon > 0$, whence

$$\lim_{x \to +\infty} \left[\frac{1}{\ln x} \int_0^x a(v) \, dv - c_1 \right] = 0.$$

Similarly, we obtain the proof of (3.11) for $x \to -\infty$.   □

Proof

Statement 1 of Theorem 3.3 Let us introduce the parameter $T > 0$ and denote for $0 \le t \le L$

$$r_T(t) = \frac{|\xi(tT)|}{\sqrt{T}}, \qquad W_T(t) = \frac{W(tT)}{\sqrt{T}},$$

$$\widehat{W}_T(t) = \int_0^t \operatorname{sign} \xi(sT) \, dW_T(s),$$

$$\beta_T(t, c_0) = \frac{1}{T} \int_0^{tT} (\xi(s)a(\xi(s)) - c_0) \, ds.$$

The process $W_T(t)$, for every fixed $T > 0$, is a Wiener process. Note that

$$\int_0^t \mathsf{P}\{\xi(s) = 0\} \, ds = 0$$

for every $t \ge 0$. According to Theorem A.4, the process $\widehat{W}_T(t)$, for every fixed $T > 0$, is also a Wiener process.

Using the Itô formula, we obtain

$$r_T^2(t) = \frac{x_0^2}{T} + (2c_0 + 1)t + 2 \int_0^t r_T(s) d\widehat{W}_T(s) + 2\beta_T(t, c_0). \tag{3.12}$$

Let us prove that

$$\lim_{T \to +\infty} \mathsf{E} \sup_{0 \le t \le L} |\beta_T(t, c_0)| = 0. \tag{3.13}$$

In order to do this, consider the function

$$\Phi(x) = 2 \int_0^x \left(\int_0^u (va(v) - c_0) \, dv \right) du.$$

According to the Itô formula, we have the equality

$$\beta_T(t, c_0) = I_T^{(1)}(t) - I_T^{(2)}(t) - I_T^{(3)}(t), \tag{3.14}$$

where

$$I_T^{(1)}(t) = \frac{\Phi(\xi(tT)) - \Phi(\xi(0))}{T}, \qquad I_T^{(2)}(t) = \frac{1}{T} \int_0^{tT} \Phi'(\xi(s)) a(\xi(s)) \, ds,$$

$$I_T^{(3)}(t) = \frac{1}{T} \int_0^{tT} \Phi'(\xi(s)) \, dW(s).$$

It is clear that

$$\frac{\Phi(x)}{x^2} \to 0, \qquad \frac{\Phi'(x)}{x} \to 0,$$

$$\Phi'(x)a(x) = \frac{\Phi'(x)}{x} (xa(x)) \to 0, \tag{3.15}$$

as $|x| \to +\infty$. Taking into account equality (3.12) we conclude that

$$\mathsf{E} \sup_{0 \le t \le L} r_T^2(t) \le C + 2\mathsf{E} \sup_{0 \le t \le L} \left| \int_0^t r_T(s) d\widehat{W}_T(s) \right|$$

$$\le C + 2 \left(\mathsf{E} \sup_{0 \le t \le L} \left| \int_0^t r_T(s) d\widehat{W}_T(s) \right|^2 \right)^{\frac{1}{2}} \le C + 2 \left(4 \int_0^L \mathsf{E} r_T^2(s) ds \right)^{\frac{1}{2}}.$$

The latter inequalities imply that

$$\mathsf{E} \sup_{0 \leq t \leq L} r_T^2(t) \leq C_L. \tag{3.16}$$

Using (3.15) we conclude that for arbitrary $\varepsilon > 0$ there exists a constant $C_\varepsilon > 0$ such that $\left| x^{-2} \Phi(x) \right| < \varepsilon$ for $|x| > C_\varepsilon$.

Therefore,

$$\mathsf{E} \sup_{0 \leq t \leq L} \left| \frac{\Phi(\xi(tT))}{T} \right| \leq \frac{C}{T} + \varepsilon \mathsf{E} \sup_{0 \leq t \leq L} r_T^2(t) \leq \frac{C}{T} + \varepsilon C_L.$$

Consequently,

$$\limsup_{T \to +\infty} \mathsf{E} \sup_{0 \leq t \leq L} \left| I_T^{(1)}(t) \right| \leq \varepsilon C_L.$$

Since the $\varepsilon > 0$ is arbitrary, we conclude that

$$\lim_{T \to +\infty} \mathsf{E} \sup_{0 \leq t \leq L} \left| I_T^{(1)}(t) \right| = 0. \tag{3.17}$$

Next, let us take arbitrary $\varepsilon > 0$ and a constant $C_\varepsilon > 0$ such that $\left| \Phi'(x) a(x) \right| < \varepsilon$ for $|x| > C_\varepsilon$. Note that

$$\mathsf{E} \sup_{0 \leq t \leq L} \left| I_T^{(2)}(t) \right| \leq \widehat{C}_\varepsilon \frac{1}{T} \int_0^{LT} \mathsf{P}\{|\xi(s)| < C_\varepsilon\} \, ds + \varepsilon L.$$

According to Lemma 3.1, we have that statement 2 of Theorem 3.1 holds. Consequently,

$$\limsup_{T \to +\infty} \mathsf{E} \sup_{0 \leq t \leq L} \left| I_T^{(2)}(t) \right| \leq \varepsilon L.$$

Therefore,

$$\mathsf{E} \sup_{0 \leq t \leq L} \left| I_T^{(2)}(t) \right| \to 0, \tag{3.18}$$

as $T \to +\infty$. Similarly to the proof of relation (3.18), we can apply the inequalities

$$\mathsf{E} \sup_{0 \leq t \leq L} \left| I_T^{(3)}(t) \right|^2 \leq \frac{4}{T^2} \int_0^{LT} \mathsf{E}\left[\Phi'(\xi(s))\right]^2 \, ds$$

$$\leq \widehat{C}_{\varepsilon} \frac{1}{T^2} \int\limits_0^{LT} \mathsf{P}\{|\xi(s)| < C_{\varepsilon}\}\, ds$$

$$+4\varepsilon^2 \int\limits_0^L \mathsf{E} r_T^2(s)\chi_{|\xi(sT)|\geq C_{\varepsilon}}\, ds \leq \widehat{C}_{\varepsilon} \frac{1}{T^2} \int\limits_0^{LT} \mathsf{P}\{|\xi(s)| < C_{\varepsilon}\}\, ds + C\varepsilon^2,$$

where

$$\widehat{C}_{\varepsilon} = \sup_x \left| \Phi'(x) \right| \chi_{|x|<C_{\varepsilon}},$$

and obtain that

$$\mathsf{E} \sup_{0\leq t\leq L} \left| I_T^{(3)}(t) \right| \to 0, \tag{3.19}$$

as $T \to +\infty$. Relations (3.17)–(3.19) imply (3.13).

Moreover, the process $\left(r_T(t), \widehat{W}_T(t), \beta_T(t, c_0)\right)$ satisfies Skorokhod's convergent subsequence principle (see Theorem A.12). According to this principle, given an arbitrary sequence $T'_n \to +\infty$ we can choose a subsequence $T_n \to +\infty$, a probability space $(\widetilde{\Omega}, \widetilde{\mathfrak{F}}, \widetilde{\mathsf{P}})$, and stochastic processes $\left(\widetilde{r}_{T_n}(t), \widetilde{W}_{T_n}(t), \widetilde{\beta}_{T_n}(r, c_0)\right)$ defined on this space such that their finite-dimensional distributions coincide with those of the processes $\left(r_{T_n}(t), W_{T_n}(t), \beta_{T_n}(r, c_0)\right)$ and, moreover, $\widetilde{r}_{T_n} \xrightarrow{\widetilde{\mathsf{P}}} \widetilde{r}(t)$, $\widetilde{W}_{T_n}(t) \xrightarrow{\widetilde{\mathsf{P}}} \widetilde{W}(t)$, $\widetilde{\beta}_{T_n}(t, c_0) \xrightarrow{\widetilde{\mathsf{P}}} 0$, as $T_n \to +\infty$, for all $0 \leq t \leq L$.

Taking into account Lemma A.13 and equality (3.12), we obtain

$$\widetilde{r}_{T_n}^2(t) = \widetilde{r}_{T_n}^2(0) + (2c_0 + 1)t + 2\int\limits_0^t \widetilde{r}_{T_n}(s)d\widetilde{W}_{T_n}(s) + 2\widetilde{\beta}_{T_n}(t, c_0). \tag{3.20}$$

Note that for arbitrary $\varepsilon > 0$

$$\widetilde{\mathsf{P}}\left\{ \left| \widetilde{r}_{T_n}(t_2) - \widetilde{r}_{T_n}(t_1) \right| > \varepsilon \right\}$$

$$\leq \widetilde{\mathsf{P}}\left\{ \left| \widetilde{r}_{T_n}(t_2) - \widetilde{r}_{T_n}(t_1) \right| > \varepsilon,\ \widetilde{r}_{T_n}(t_2) + \widetilde{r}_{T_n}(t_1) > \varepsilon \right\}$$

$$\leq \widetilde{\mathsf{P}}\left\{ \left| \widetilde{r}_{T_n}^2(t_2) - \widetilde{r}_{T_n}^2(t_1) \right| > \varepsilon^2 \right\}. \tag{3.21}$$

Taking into account (3.20), we get

$$\lim_{h\to 0} \limsup_{T_n\to+\infty} \sup_{|t_1-t_2|\leq h,\ t_i\leq L} \widetilde{\mathsf{P}}\left\{ \left| \widetilde{r}_{T_n}(t_2) - \widetilde{r}_{T_n}(t_1) \right| > \varepsilon \right\} = 0.$$

Therefore, according to Lemma A.7, we can pass to the limit in the stochastic integral in (3.20). Thus,

$$\tilde{r}^2(t) = (2c_0 + 1)t + 2\int_0^t \tilde{r}(s)d\widetilde{W}(s).$$

Since the subsequence T_n is arbitrary and the solution of the latter equation is unique, the finite-dimensional distributions of the process $r_T(t)$ tend, as $T \to +\infty$, to the corresponding finite-dimensional distributions of the process $r(t)$.

Applying inequalities (2.21), we get that for arbitrary $\varepsilon > 0$

$$\mathsf{P}\left\{\sup_{|t_1-t_2|\leq h,\ t_i\leq L}\left|\int_{t_1}^{t_2} r_T(s)\,d\widehat{W}_T(s)\right| > \varepsilon\right\}$$

$$\leq \mathsf{P}\left\{4\sup_{kh<L}\sup_{kh\leq t\leq(k+1)h}\left|\int_{kh}^{t} r_T(s)d\widehat{W}_T(s)\right| > \varepsilon\right\}$$

$$\leq \sum_{kh<L}\mathsf{P}\left\{\sup_{kh\leq t\leq(k+1)h}\left|\int_{kh}^{t} r_T(s)d\widehat{W}_T(s)\right| > \frac{\varepsilon}{4}\right\}$$

$$\leq \sum_{kh<L}\left(\frac{4}{\varepsilon}\right)^4\mathsf{E}\left|\int_{kh}^{(k+1)h} r_T(s)d\widehat{W}_T(s)\right|^4 \leq \sum_{kh<L}\left(\frac{4}{\varepsilon}\right)^4 6h\int_{kh}^{(k+1)h}\mathsf{E}r_T^4(s)ds.$$

Taking into account (3.12), we obtain the inequality

$$\mathsf{E}r_T^4(t) \leq C.$$

The latter inequalities imply that

$$\lim_{h\to 0}\limsup_{T\to+\infty}\mathsf{P}\left\{\sup_{|t_1-t_2|\leq h,\ t_i\leq L}\left|r_T^2(t_2) - r_T^2(t_1)\right| > \varepsilon\right\} = 0.$$

Taking into account (3.21), we have the relation

$$\lim_{h\to 0}\limsup_{T\to+\infty}\mathsf{P}\left\{\sup_{|t_1-t_2|\leq h,\ t_i\leq L}\left|r_T(t_2) - r_T(t_1)\right| > \varepsilon\right\} = 0$$

for any $t_i \leq L, \varepsilon > 0, L > 0$.

According to Theorem A.13 the stochastic process $r_T(t)$ converges weakly, as $T \to +\infty$, to the solution $r(t)$ of Eq. (3.10) with $c = c_0$.

Statement 2 Let us take $\varepsilon > 0$ such that $2c_1 - 2\varepsilon > 1$ and $2c_2 + 2\varepsilon < 1$. Using Lemma 3.1 and equality (3.4) with $c_0 = c_1$ for $x > L$ and $c_0 = c_2$ for $x < -L$, we obtain the following inequalities

$$f'(x) \leq x^{-2c_1} x^{2\varepsilon} \quad \text{for } x > L$$

and

$$f'(x) \geq |x|^{-2c_2} |x|^{-2\varepsilon} \quad \text{for } x < -L.$$

So, the function $f(x)$ is bounded from above and $f(x) \to -\infty$, as $x \to -\infty$. Thus (see Lemma A.5),

$$P\left\{ \lim_{t \to +\infty} \xi(t) = +\infty \right\} = 1. \tag{3.22}$$

Now let us use the analog of equality (3.12), that is

$$r_T^2(t) = \frac{x_0^2}{T} + \frac{1}{T} \int_0^{tT} [2\bar{c}(\xi(s)) + 1] \, ds + 2 \int_0^t r_T(s) d\widehat{W}_T(s) + 2\beta_T(t, \bar{c}(\xi(s))) ,$$

where

$$\beta_T(t, \bar{c}(\xi(t))) = \frac{1}{T} \int_0^{tT} [\xi(s)a(\xi(s)) - \bar{c}(\xi(s))] \, ds.$$

According to the Itô formula for $\beta_T(t, \bar{c}(\xi(t)))$, we have the analog of equality (3.14), where

$$\Phi(x) = 2 \int_0^x \left(\int_0^u (va(v) - \bar{c}(v)) \, dv \right) du.$$

Completely analogous to the proof of (3.13), we obtain

$$\lim_{T \to +\infty} \mathsf{E} \sup_{0 \leq t \leq L} |\beta_T(t, \bar{c}(\xi(t)))| = 0.$$

The rest of the proof can be done in the same way as in the proof of statement 1. In doing so use (3.22) and the convergence for $t \geq 0$: $\xi(tT)T^{-\frac{1}{2}} - |\xi(tT)| T^{-\frac{1}{2}} \to 0$, as $T \to +\infty$, with probability one, that follows from (3.22).

The proof of statement 3 is literally the same as that of statement 2 with the only difference that the function $f(x)$ is bounded from below and $f(x) \to +\infty$, as $x \to +\infty$. In this case (see Lemma A.6)

$$P\left\{ \lim_{t \to +\infty} \xi(t) = -\infty \right\} = 1. \tag{3.23}$$

$\square$

Remark 3.3 Let the relation

$$xa(x) - \overline{c}(x) \to 0,$$

as $|x| \to +\infty$, hold. Then we have (3.9).

In fact, it follows from the equality

$$\frac{1}{x} \int_0^x va(v)\, dv - \overline{c}(x) = \frac{1}{x} \int_0^x [va(v) - \overline{c}(v)]\, dv.$$

Corollary 3.1 *For $2c - 1 \geq 0$ there exists a unique strong solution $r(t)$ to SDE*

$$d\,r(t) = \frac{c}{r(t)} dt + dW(t). \tag{3.24}$$

Indeed, the proof of Theorem 3.3 implies the existence of a unique process $r(t)$, which is a strong solution to Eq. (3.10). According to Remark 3.3, for $2c - 1 \geq 0$ the origin is not attainable by the process $r(t)$. Therefore, we can apply the Itô formula to the process $\Phi\left(r^2(t)\right)$, where $\Phi(x) = \sqrt{|x|}$, and obtain that the process $r(t)$ satisfies Eq. (3.24).

Next, we show that, under additional conditions on the convergence rate in (3.9), we obtain equalities instead of inequalities for c_1 and c_2 in Theorem 3.3. That is, the following theorem holds.

Theorem 3.4 *Let the assumptions of Theorem 3.3 be fulfilled.*

1. *If $c_1 = c_2 = -\frac{1}{2}$ and*

$$\lim_{|x| \to +\infty} \ln|x|\, \psi\left(x, -\frac{1}{2}\right) = +\infty,$$

then the stochastic process $T^{-\frac{1}{2}}\xi(tT)$ converges weakly, as $T \to +\infty$, to the process $r(t) \equiv 0$.

2. *If*

   (1) $2c_1 > 1$, $2c_2 = 1$, *and there exist* $\varepsilon(x) > 0$ *and constants* $C > 0$ *and* $L > 0$
       such that

   $$\int_0^{+\infty} \frac{\varepsilon(x)}{x} dx = +\infty \quad and \quad \ln \varepsilon(x) + 2 \ln |x| \, \psi \left(x, \frac{1}{2} \right) < C \qquad (3.25)$$

   for $x < -L$,
   or
   (2) $2c_1 = 1$, $2c_2 = 1$ *and, in addition to the assumptions (3.25), there exists a*
       function $\varepsilon_1(x) > 0$ *such that*

   $$\int_0^{+\infty} \frac{\varepsilon_1(x)}{x} dx < +\infty \quad and \quad \ln \varepsilon_1(x) + 2 \ln |x| \, \psi \left(x, \frac{1}{2} \right) > -C$$

   for $x > L$, *then the stochastic process* $T^{-\frac{1}{2}} \xi(tT)$ *converges weakly, as* $T \to$
   $+\infty$, *to the solution* $r(t)$ *of Eq. (3.10) with* $c = c_1$.

3. *If*

   (1) $2c_1 = 1$, $2c_2 > 1$ *and there exist* $\varepsilon(x) > 0$ *and constants* $C > 0$ *and* $L > 0$
       such that

   $$\int_0^{+\infty} \frac{\varepsilon(x)}{x} dx = +\infty \quad and \quad \ln \varepsilon(x) + 2 \ln |x| \, \psi \left(x, \frac{1}{2} \right) < C \qquad (3.26)$$

   for $x > L$,
   or
   (2) $2c_1 = 1$, $2c_2 = 1$ *and, in addition to the assumptions (3.26), there exists a*
       function $\varepsilon_1(x) > 0$ *such that*

   $$\int_0^{-\infty} \frac{\varepsilon_1(x)}{x} dx < +\infty \quad and \quad \ln \varepsilon_1(x) + 2 \ln |x| \, \psi \left(x, \frac{1}{2} \right) > -C$$

   for $x < -L$, *then the stochastic process* $-T^{-\frac{1}{2}} \xi(tT)$ *converges weakly, as*
   $T \to +\infty$, *to the solution* $r(t)$ *of Eq. (3.10) with* $c = c_2$.

Proof Assumption (1) of Theorem 3.4 implies the Statement 2 of Theorem 3.2.
Furthermore, the proof of assertion (1) of Theorem 3.3 implies that assumption (1)
of Theorem 3.4 is sufficient for the weak convergence of the process $T^{-\frac{1}{2}} \xi(tT)$, as

$T \to +\infty$, to the solution $r(t)$ of Eq. (3.10) with $2c + 1 = 0$. From the uniqueness of the solution of Eq. (3.10) we have that $r(t) \equiv 0$.

Assumption (2) of Theorem 3.4 implies the relations $f(x) < C$ and $f(x) \to -\infty$, as $x \to -\infty$, that is, relation (3.22) holds. The proof of assertion (2) of Theorem 3.3 implies that statement (2) of Theorem 3.4 is sufficient for the weak convergence of the process $T^{-\frac{1}{2}}\xi(tT)$, as $T \to +\infty$, to the solution $r(t)$ of Eq. (3.10) with $c = c_1$.

In the proof of statement (3) of Theorem 3.4 we use relation (3.23), which follows from the fact that in this case $f(x) \to +\infty$, as $x \to +\infty$, and the function $f(x)$ is bounded from below.

The theorem is proved. $\square$

3.3   Influence of the Coefficients of the Equation on the Limit Behavior of the Solutions

The results obtained in the process of studying of Eq. (3.1) allow us to investigate the asymptotic behavior of the solutions to a class of equations of the form

$$d\eta(t) = a\left(\eta(t)\right)dt + \sigma\left(\eta(t)\right)dW(t), \quad \eta(0) = x_0, \tag{3.27}$$

where the function $\sigma(x) > 0$ is continuously differentiable, and $g(x) \to -\infty$, as $x \to -\infty$, $g(x) \to +\infty$, as $x \to +\infty$, where $g(x) = \int_0^x \frac{dy}{\sigma(y)}$.

In fact, let us consider the process $\xi(t) = g\left(\eta(t)\right)$. According to the Itô formula, we have the equality

$$d\xi(t) = \widehat{a}\left(\xi(t)\right)dt + dW(t), \tag{3.28}$$

where

$$\widehat{a}(x) = \frac{a\left(l(x)\right)}{\sigma\left(l(x)\right)} - \frac{1}{2}\sigma'\left(l(x)\right),$$

here the function $l(x)$ is the inverse function to $g(x)$. So, we obtain that ξ is the solution of Eq. (3.1).

In this case $g'(x) = \frac{1}{\sigma(x)} > 0$ for all $x \in \mathbb{R}$. Consequently, the function $g(x)$ is strictly monotonously increasing. The equality $g\left(l(x)\right) = x$, that holds for all $x \in \mathbb{R}$, implies the convergences $l(x) =\to -\infty$, as $x \to -\infty$, and $l(x) =\to +\infty$, as $x \to +\infty$. Since $l'(x) = \frac{1}{g'(l(x))} > 0$ the inverse function is strictly monotonously increasing.

Let us formulate the analogs of Theorems 3.1 and 3.3.

Theorem 3.5 *Let η be a solution of Eq. (3.27) and let*

$$\lim_{|x|\to+\infty} \frac{1}{\ln|x|} \int_0^{l(x)} \left[\frac{a(z)}{\sigma^2(z)} - \frac{1}{2}\frac{\sigma'(z)}{\sigma(z)} \right] dz = c_0.$$

Then we have two cases.

1. *If $2c_0 < -1$, then the stochastic process η is ergodic and*

$$\lim_{t\to+\infty} \mathsf{P}\{\eta(t) < x\} = \left[\int_{\mathbb{R}} \frac{dz}{f'(z)} \right]^{-1} \int_{-\infty}^{g(x)} \frac{dz}{f'(z)}, \qquad (3.29)$$

where

$$f'(x) = \exp\left\{ -2 \int_0^{l(x)} \left[\frac{a(v)}{\sigma^2(v)} - \frac{1}{2}\frac{\sigma'(v)}{\sigma(v)} \right] dv \right\}.$$

2. *If $2c_0 > -1$ and $|x\widehat{a}(x)| \le C$, then the stochastic process η is stochastically unstable, in other words*

$$\lim_{t\to+\infty} \frac{1}{t} \int_0^t \mathsf{P}\{|\eta(s)| < N\}\, ds = 0 \qquad (3.30)$$

for any constant $N > 0$.

Proof In this case, all the conditions of Theorem 3.1 are satisfied for Eq. (3.28) with the given c_0. Therefore, if $2c_0 < -1$, then we have relation (3.29) for the process η, and if $2c_0 > -1$, then relation (3.30) holds. □

Theorem 3.6 *Let η be a solution to Eq. (3.27) and let $|x\widehat{a}(x)| \le C$. If*

$$\lim_{|x|\to+\infty} \left\{ \frac{1}{g(x)} \int_0^x g(y) \left[\frac{a(y)}{\sigma^2(y)} - \frac{1}{2}\frac{\sigma'(y)}{\sigma(y)} \right] dy - \overline{c}(x) \right\} = 0,$$

where

$$\overline{c}(x) = \begin{cases} c_1, & x \ge 0, \\ c_2, & x < 0, \end{cases}$$

then the stochastic process $\xi(t) = g(\eta(t))$ satisfies statements (1)–(3) of Theorem 3.3.

Proof It is easy to get that

$$\frac{1}{x} \int_0^x v\widehat{a}(v)\, dv = \frac{1}{x} \int_0^{l(x)} g(y) \left[\frac{a(y)}{\sigma^2(y)} - \frac{1}{2} \frac{\sigma'(y)}{\sigma(y)} \right] dy,$$

$$\overline{c}(x) = \overline{c}(g(x)).$$

Therefore, the conditions of Theorem 3.3 are satisfied for the coefficients of Eq. (3.28). So, Theorem 3.3 implies Theorem 3.6.   □

Similarly, we can obtain analogs of Theorems 3.2 and 3.4 for the process $g(\eta(t))$, where $\eta(t)$ is the solution to Eq. (3.27).

Note that the analogs of Theorems 3.3 and 3.4 assert the weak convergence of the stochastic process $T^{-\frac{1}{2}} |g(\eta(tT))|$, as $T \to +\infty$, to the process $r(t)$. If additionally we have the convergence

$$\frac{1}{x} \int_0^x \frac{dv}{\sigma(v)} \to \sigma_0, \quad \sigma_0 > 0,$$

as $|x| \to +\infty$, then we obtain the weak convergence of the stochastic process $|\eta(tT)| T^{-\frac{1}{2}}$ to the process $\sigma_0^{-1} r(t)$.

3.4   Influence of the Diffusion Coefficient on the Limit Behavior of the Solutions

Now, we consider the asymptotic behavior, as $t \to +\infty$, of the distributions of the solutions ξ of Eq. (2.1), in which the drift coefficients $a(x) \equiv 0$. More precisely, we consider equations of the form

$$d\xi(t) = \sigma(\xi(t))\, dW(t), \quad \xi(0) = x_0, \tag{3.31}$$

where $\sigma(x) \sim c|x|^\alpha$, as $|x| \to +\infty$, $0 < \alpha < \frac{1}{2}$, and $\sigma(x) > 0$ is a continuously differentiable function.

Note that for $\alpha > \frac{1}{2}$ the solution ξ to Eq. (3.31) is an ergodic process (see Theorem A.10), and

$$\lim_{t \to +\infty} P\{\xi(t) < x\} = \left[\int_{\mathbb{R}} \frac{dv}{\sigma^2(v)} \right]^{-1} \int_{-\infty}^x \frac{dv}{\sigma^2(v)}.$$

Theorem 3.7 *Let ξ be a solution to Eq. (3.31). If for all x*

$$\sigma(x) = c_0|x|^\alpha + \beta(x),$$

where $c_0 > 0$, $0 < \alpha < \frac{1}{2}$, $\beta(x) = o(|x|^\alpha)$, $\beta'(x) = o(|x|^{\alpha-1})$, as $|x| \to +\infty$, then the finite-dimensional distributions of the process $T^{-\frac{1}{2}(1-\alpha)}|\xi(tT)|$ converge, as $T \to +\infty$, to the corresponding distributions of the process $[c_0(1-\alpha)r(t)]^{\frac{1}{1-\alpha}}$, where $r(t)$ is a solution to Eq. (3.10) with $c = -\frac{\alpha}{2(1-\alpha)}$.

Proof The function $g(x) = \int\limits_0^x \frac{dv}{\sigma(v)}$ is increasing. It is twice continuously differentiable and $g(x) \to +\infty$, as $x \to +\infty$, $g(x) \to -\infty$ as $x \to -\infty$. It is clear that

$$\frac{g(x)c_0(1-\alpha)}{|x|^{1-\alpha}\operatorname{sign} x} \to 1, \qquad \frac{l(x)}{[|x|c_0(1-\alpha)]^{\frac{1}{1-\alpha}}\operatorname{sign} x} \to 1, \qquad (3.32)$$

as $|x| \to +\infty$, where the function $l(x)$ is the inverse function to $g(x)$. We can apply the Itô formula to the process $\eta(t) = g(\xi(t))$ and obtain the equation

$$d\eta(t) = \widehat{a}(\eta(t))dt + dW(t), \qquad (3.33)$$

where $\widehat{a}(x) = -\frac{1}{2}\sigma'(l(x))$. Obvious that

$$x\widehat{a}(x) \to -\frac{\alpha}{2(1-\alpha)},$$

as $|x| \to +\infty$. In addition,

$$-\frac{\alpha}{1-\alpha} + 1 = \frac{1-2\alpha}{1-\alpha} > 0.$$

Consequently, all the conditions of statement 1 of Theorem 3.3 are fulfilled for Eq. (3.33). Therefore, the process $r_T(t) = T^{-\frac{1}{2}}|\xi(tT)|$ converges weakly, as $T \to +\infty$, to the solution $r(t)$ of Eq. (3.10) with $c = -\frac{\alpha}{2(1-\alpha)}$. According to relation (3.32), for arbitrary $\varepsilon > 0$ there exists a constant C_ε such that for $|x| > C_\varepsilon$ we have the inequality

$$\left| |g(x)|c_0(1-\alpha)|x|^{\alpha-1} - 1 \right| < \varepsilon.$$

Therefore,

$$|r_T(t) - \xi_T(t)| \leq \frac{1}{\sqrt{T}} \sup_{|x| \leq c_\varepsilon} \left| |g(x)| - |x|^{1-\alpha}[c_0(1-\alpha)|]^{-1} \right| + \varepsilon\xi_T(t), \qquad (3.34)$$

where

$$\xi_T(t) = \left[\sqrt{T}\, c_0(1 - \alpha)\right]^{-1} |\xi(tT)|^{1-\alpha}.$$

Using the boundedness of the function $x\widehat{a}(x)$, and taking into account (3.33) as well as the inequality

$$r_T(t) = \frac{|g(\xi(tT))|\, c_0(1 - \alpha)}{|\xi(tT)|^{1-\alpha}} \xi_T(t) \geq \delta\, \xi_T(t),$$

we conclude that

$$\mathsf{E}\xi_T(t) \leq C + C_1 t. \tag{3.35}$$

Since $\varepsilon > 0$ is arbitrary, it follows from (3.34) and (3.35) that

$$r_T(t) - \xi_T(t) \to 0$$

in probability, as $T \to +\infty$.

So, the finite-dimensional distributions of the process $\xi_T(t)$ converge, as $T \to +\infty$, to the corresponding distributions of the solution $r(t)$ of Eq. (3.10) with $c = -\frac{\alpha}{2(1-\alpha)}$. Therefore, the finite-dimensional distributions of the process

$$T^{-\frac{1}{2(1-\alpha)}} |\xi(tT)|$$

converge, as $T \to +\infty$, to the corresponding distributions of the process

$$[c_0(1 - \alpha)r(t)]^{\frac{1}{1-\alpha}}.$$

$\square$

Remark 3.4 Note that the case $\alpha = \frac{1}{2}$ is critical. In particular, if we add to Eq. (3.31) the drift coefficient $a(x) = -\frac{1}{2}\operatorname{sign} x$, then the solution ξ to such an equation has an ergodic distribution. This fact is an immediate consequence of the following theorem.

Theorem 3.8 *Let $\eta(t)$ be a solution to Eq. (3.27). If*

$$\lim_{|x| \to +\infty} \frac{2xa(x)}{\sigma^2(x)} = -1, \quad \sigma(x) = \begin{cases} c_1|x|^{\alpha_1} + \beta_1(x), & x \geq 0, \\ c_2|x|^{\alpha_2} + \beta_2(x), & x < 0, \end{cases}$$

where $c_i > 0$, $0 < \alpha_i \leq 1$, $\beta_i(x) = o(|x|^{\alpha_i})$, as $|x| \to +\infty$, then

$$\lim_{t \to +\infty} \mathsf{P}\{\eta(t) < x\} = B^{-1} \int_{-\infty}^{x} \frac{du}{f'(u)\sigma^2(u)},$$

where

$$f'(x) = \exp\left\{-2\int_0^x \frac{a(u)}{\sigma^2(u)}du\right\}, \quad B = \int_{\mathbb{R}} \frac{du}{f'(u)\sigma^2(u)}.$$

Proof According to the Itô formula for the process $\xi(t) = f(\eta(t))$, where $f(x) = \int_0^x f'(u)du$, we obtain the equation

$$\xi(t) = f(\eta(0)) + \int_0^t \widehat{\sigma}(\xi(s))dW(s),$$

where $\widehat{\sigma}(x) = f'(\varphi(x))\,\sigma\,(\varphi(x))$ and the function $\varphi(x)$ is the inverse function to $f(x)$.

Since

$$\frac{2xa(x)}{\sigma^2(x)} \to -1,$$

as $|x| \to +\infty$, for arbitrary $\varepsilon > 0$ there exists a constant C_ε such that for $|x| > C_\varepsilon$ we have the inequalities

$$-1 - \varepsilon < \frac{2xa(x)}{\sigma^2(x)} < -1 + \varepsilon.$$

Consider $f'(x)$, and let $x > 0$. For $x > C_\varepsilon$

$$f'(x) = \exp\left\{-2\int_0^{C_\varepsilon} \frac{a(u)}{\sigma^2(u)}du - 2\int_{C_\varepsilon}^x \frac{a(u)}{\sigma^2(u)}du\right\} > \widetilde{K}_\varepsilon \exp\left\{\int_{C_\varepsilon}^x \frac{-2ua(u)}{u\sigma^2(u)}du\right\}$$

$$> \widetilde{K}_\varepsilon \exp\left\{(1 - \varepsilon)\ln\frac{x}{C_\varepsilon}\right\} = \widetilde{K}_\varepsilon \left(\frac{x}{C_\varepsilon}\right)^{1-\varepsilon} = K_\varepsilon x^{1-\varepsilon}.$$

The similar situation is for $x < 0$.

So, we obtain the inequality $f'(x) > K_\varepsilon |x|^{1-\varepsilon}$. Consequently,

$$f(+\infty) = +\infty, \quad f(-\infty) = -\infty$$

and

$$\int_{\mathbb{R}} \frac{du}{\widehat{\sigma}^2(u)} < +\infty.$$

According to Theorem A.10, we obtain

$$P\{\eta(t) < x\} = P\{\xi(t) < f(x)\} \rightarrow \left[\int_{\mathbb{R}} \frac{du}{\widehat{\sigma}^2(u)}\right]^{-1} \int_{-\infty}^{f(x)} \frac{du}{\widehat{\sigma}^2(u)},$$

as $t \rightarrow +\infty$. Hence the proof of Theorem 3.8 follows.   □

3.5 Examples

Consider the following examples of the drift coefficient $a(x)$ in Eq. (3.1).

Example 3.1 Let

$$a(x) = \bar{c}(x)\frac{x}{1+x^2} + \frac{x \sin x}{1+x^2}, \quad \bar{c}(x) = \begin{cases} c_1, & x > 0, \\ c_2, & x < 0, \end{cases} \quad \sigma(x) = 1.$$

Since for all x we have $|xa(x)| \leq C$ and

$$\frac{1}{x}\int_0^x va(v)dv = \frac{1}{x}\int_0^x \bar{c}(v)\frac{v^2}{1+v^2}dv + \frac{1}{x}\int_0^x \frac{v^2 \sin v}{1+v^2}dv$$

$$= \bar{c}(x) + \frac{1}{x}\int_0^x \bar{c}(v)\left[\frac{v^2}{1+v^2} - 1\right]dv + \frac{1}{x}\int_0^x \sin v\, dv - \frac{1}{x}\int_0^x \frac{\sin v}{1+v^2}dv \rightarrow \bar{c}(x),$$

as $|x| \rightarrow +\infty$, then

$$\lim_{|x|\to+\infty}\left[\frac{1}{x}\int_0^x va(v)dv - \bar{c}(x)\right] = 0.$$

According to Theorem 3.3 we obtain:

(1) if $c_1 = c_2 = c_0$ and $2c_0 > -1$, then the stochastic process $|\xi(tT)|\,T^{-1}$ converges weakly, as $T \rightarrow +\infty$, to the Bessel diffusion process $r(t)$, which is the solution to Eq. (3.10) for $c = c_0$;

(2) if $2c_1 > 1$, $2c_2 < 1$, then the stochastic process $\xi(tT)T^{-\frac{1}{2}}$ converges weakly, as $T \rightarrow +\infty$, to the Bessel diffusion process $r(t)$, which is the solution to Eq. (3.10) for $c = c_1$;

(3) if $2c_1 < 1$, $2c_2 > 1$, then the stochastic process $-\xi(tT)T^{-\frac{1}{2}}$ converges weakly, as $T \to +\infty$, to the Bessel diffusion process $r(t)$, which is the solution to Eq. (3.10) for $c = c_2$.

If $c_1 = c_2 = c_0$ and $2c_0 < -1$, then, according to Theorem 3.1, the stochastic process $\xi(t)$ is ergodic and

$$P\{\xi(t) < x\} \to \left[\int_{\mathbb{R}} \frac{dv}{f'(v)}\right]^{-1} \int_{-\infty}^{x} \frac{dv}{f'(v)},$$

as $t \to +\infty$, where $f'(x) = e^{-2\int_0^x a(v)dv}$.

Consequently, the ergodicity and instability of the solution depend on the magnitude c_0.

Example 3.2 Let

$$a(x) = -\frac{1}{2}\frac{x}{2+x^2} - 2\frac{x}{(2+x^2)\ln(2+x^2)}, \quad \sigma(x) = 1.$$

Since

$$\int_0^x a(v)dv = -\frac{1}{4}\ln(2+x^2) + \frac{1}{4}\ln 2 - \ln\,\ln(2+x^2) + \ln\,\ln 2,$$

we have

$$f'(x) = \exp\left\{-2\int_0^x a(v)\,dv\right\} = \sqrt{2+x^2}\left[\ln(2+x^2)\right]^2 \cdot c_1,$$

where

$$c_1 = e^{-\frac{1}{2}\ln 2 - 2\ln\,\ln 2}.$$

Taking into account that

$$\int_{\mathbb{R}} \frac{dv}{\sqrt{2+v^2}\left[\ln(2+v^2)\right]^2} = 2\int_0^{+\infty} \frac{dv}{\sqrt{2+v^2}\left[\ln(2+v^2)\right]^2} = \int_2^{+\infty} \frac{dz}{\sqrt{z}\sqrt{z-2}\,(\ln z)^2} < +\infty,$$

we obtain the inequality

$$\int_{\mathbb{R}} \frac{dx}{f'(x)} < +\infty.$$

Consequently, according to Theorem 3.2, the solution ξ to Eq. (3.1) is ergodic, and

$$P\{\xi(t) < x\} \to \left[\int_{\mathbb{R}} \frac{dv}{f'(v)}\right]^{-1} \int_{-\infty}^{x} \frac{dv}{f'(v)},$$

as $t \to +\infty$.

Example 3.3 Let

$$a(x) = -\frac{1}{2}\frac{x}{1+x^2} + \frac{2x}{(2+x^2)\ln(2+x^2)}, \quad \sigma(x) = 1.$$

$$\int_{0}^{x} a(v)dv = -\frac{1}{4}\ln(2+x^2) + \frac{1}{4}\ln 2 + \ln \ln(2+x^2) - \ln \ln 2.$$

Then

$$f'(x) = \exp\left\{-2\int_{0}^{x} a(v)\,dv\right\} = \sqrt{2+x^2}\left[\ln(2+x^2)\right]^{-2} \cdot c_2,$$

where

$$c_2 = e^{-\frac{1}{2}\ln 2 + 2\ln \ln 2}.$$

Thus,

$$\frac{f'(x)}{x} = \frac{\sqrt{2+x^2}}{x} \cdot \frac{1}{\left[\ln(2+x^2)\right]^2} \to 0,$$

as $|x| \to +\infty$. Consequently, according to Theorem 3.2, the solution ξ to Eq. (3.1) is stochastically unstable, that is, for an arbitrary constant $N > 0$

$$\lim_{t \to +\infty} \frac{1}{t}\int_{0}^{t} P\{|\xi(s)| < N\}\,ds = 0.$$

In addition, in this case $c_1 = c_2 = -\frac{1}{2}$ and

$$\lim_{|x| \to +\infty} \ln|x|\,\psi\left(x, -\frac{1}{2}\right) = \lim_{|x| \to +\infty} \ln\left\{\frac{\sqrt[4]{|x|}}{\sqrt[4]{2+x^2}}\sqrt{\ln(2+x^2)}\right\} + \ln\frac{\sqrt[4]{2}}{\sqrt{\ln 2}} = +\infty.$$

According to Theorem 3.4, the stochastic process $\xi(tT)T^{-\frac{1}{2}}$ converges weakly, as $T \to +\infty$ to the process $r(t) \equiv 0$.

Consider the following examples of the diffusion coefficient $\sigma(x)$ in Eq. (3.31).

Example 3.4 Let $\sigma(x) = \sqrt[6]{1+x^2}$.

Thus, $\sigma(x) \sim |x|^{\frac{1}{3}}$, as $|x| \to +\infty$. Therefore, according to Theorem 3.7 with $c_0 = 1$ and $\alpha = \frac{1}{3}$, the finite-dimensional distributions of the process $|\xi(tT)| T^{-\frac{1}{3}}$ converge, as $T \to +\infty$, to the corresponding distributions of the process $\left(\frac{2}{3}r(t)\right)^{\frac{3}{2}}$, where $r(t)$ is the solution of Eq. (3.10) for $c = -\frac{1}{4}$.

Example 3.5 Let

$$\sigma(x) = \frac{1}{\sqrt[6]{1+x^2}}.$$

Thus, $\sigma(x) \sim |x|^{-\frac{1}{3}}$, as $|x| \to +\infty$. Therefore, according to Theorem 3.7 with $c_0 = 1$ and $\alpha = -\frac{1}{3}$, the finite-dimensional distributions of the process $|\xi(tT)| T^{-\frac{2}{3}}$ converge, as $T \to +\infty$, to the corresponding distributions of the process $\left(\frac{4}{3}r(t)\right)^{\frac{3}{4}}$, where $r(t)$ is the solution to Eq. (3.10) for $c = \frac{1}{8}$.

Chapter 4
Asymptotic Behavior of Integral Functionals of Stochastically Unstable Solutions

A very important class of functionals from solutions ξ to SDEs is represented by stochastic integrals

$$\int_0^t g\left(\xi\left(s\right)\right) ds \text{ and } \int_0^t g\left(\xi\left(s\right)\right) d\xi\left(s\right),$$

where $g = g(x)$ is a non-random function. It is supposed that ξ is stochastically unstable solution of some Itô's SDE, and the integrals exist for every $t > 0$ in the respective sense. Also, we assume that the integrals are unbounded in probability, as $t \to +\infty$.

In this chapter we study the behavior of the distributions, as $t \to +\infty$, of these functionals after some normalization. For example, according to the Itô formula, we have with probability 1 the equality

$$1 - \cos W(t) = \frac{1}{2} \int_0^t \cos W(s)\, ds + \int_0^t \sin W(s)\, dW(s), \tag{4.1}$$

for all $t \geq 0$, where W is a standard Wiener process. It is clear that W is stochastically unstable solution of SDE (2.1) for $a(x) \equiv 0$, $\sigma(x) \equiv 1$, and $\xi(0) = 0$. We know that (see Example 4.2) the stochastic process

$$\beta_T^{(2)}(t) = \frac{1}{\sqrt{T}} \int_0^{tT} \sin W(s)\, dW(s), \quad t \geq 0$$

© Springer Nature Switzerland AG 2020
G. Kulinich et al., *Asymptotic Analysis of Unstable Solutions of Stochastic Differential Equations*, Bocconi & Springer Series 9,
https://doi.org/10.1007/978-3-030-41291-3_4

converges weakly, as $T \rightarrow +\infty$, to the process $\frac{1}{\sqrt{2}} W^*(t)$, where W^* is a Wiener process. So the functional $\int_0^t \sin W(s) \, dW(s)$, $t > 0$ is unbounded in probability, the equality (4.1) imply that the functional $\int_0^t \cos W(s) \, ds$ is unbounded in probability as well. Since $\frac{1}{\sqrt{T}} \cos W(Tt) \rightarrow 0$ with probability 1, as $T \rightarrow +\infty$, for all $t \geq 0$, the stochastic processes

$$\beta_T^{(1)}(t) = \frac{1}{\sqrt{T}} \int_0^{tT} \cos W(s) \, ds$$

converge weakly, as $T \rightarrow +\infty$, to the process $-\sqrt{2} W^*(t)$, where W^* is a Wiener process. It should be emphasized that these convergences are similar, in some sense, to the central limit theorem for the sum of the dependent random variables.

We note that during the study of asymptotic behavior, as $t \rightarrow +\infty$, of the distributions of unstable solutions ξ of Eq. (2.1), which is equivalent to the integral equation

$$\xi(t) = \xi(0) + \int_0^t a(\xi(s)) \, ds + \int_0^t \sigma(\xi(s)) \, dW(s), \tag{4.2}$$

we investigate the asymptotic behavior, as $t \rightarrow +\infty$, of the distributions of the integral functionals, that are unbounded in probability.

Therefore, in the study of the asymptotic behavior, as $t \rightarrow +\infty$, of the distributions of stochastically unbounded integral functionals two problems arise: the problem of finding appropriate non-random normalizing multipliers $B(t) \rightarrow +\infty$, as $t \rightarrow +\infty$, that ensure convergence of the distributions of normalized functionals to the distributions of some non-degenerate distributions, and the problem of describing the class of limit distributions.

Note that

$$\int_0^t g(\xi(s)) \, d\xi(s) = \int_0^t g(\xi(s)) a(\xi(s)) \, ds + \int_0^t g(\xi(s)) \sigma(\xi(s)) \, dW(s),$$

so in this chapter we study the asymptotic behavior of the distributions, as $T \rightarrow +\infty$, of the following functionals:

$$\beta_T^{(1)}(t) = \frac{1}{B_1(T)} \int_0^{tT} g(\xi(s)) \, ds, \quad \beta_T^{(2)}(t) = \frac{1}{B_2(T)} \int_0^{tT} g(\xi(s)) \, dW(s), \quad t \geq 0,$$

where $B_i(T)$ are non-random normalizing multipliers and $B_i(T) \to +\infty$, as $T \to +\infty$, $g(x)$ is non-random measurable locally integrable (for $\beta_T^{(1)}(t)$) or locally square integrable (for $\beta_T^{(2)}(t)$) function, the stochastic processes ξ and W are related via Eq. (2.1) or via Eq. (3.1).

The chapter consists of three sections. In Sect. 4.1 the weak convergence is studied, as $T \to +\infty$, for the functionals $\beta_T^{(1)}(t)$ and $\beta_T^{(2)}(t)$ of unstable solutions of Eq. (2.1). In Theorem 4.1 the sufficient conditions are established for the weak convergence, as $T \to +\infty$, of the functional $\beta_T^{(1)}(t)$ to the process

$$
\beta^{(1)}(t) = 2 \left[\int_0^{\zeta(t)} |u|^\alpha \, \overline{b}(u) \, du - \int_0^t |\zeta(s)|^\alpha \, \overline{b}(\zeta(s)) \, d\zeta(s) \right],
$$

where ζ is the solution to Eq. (2.8), $\alpha \geq 0$ is the order of regularity at infinity of the function ψ (see Definition 4.1 in Sect. 4.1),

$$
\overline{b}(x) = \begin{cases} b_1, & x \geq 0, \\ b_2, & x < 0, \end{cases}
$$

b_1 and b_2 are some constants. In Theorem 4.2 the sufficient conditions are established for the weak convergence, as $T \to +\infty$, of the functional $\beta_T^{(2)}(t)$ to the process $\beta^{(2)}(t) = W^*\left(\beta^{(1)}(t)\right)$, where the process $\beta^{(1)}(t)$ has form (4.4), $W^*(t)$ is a Wiener process, the processes $W^*(t)$ and $\beta^{(1)}(t)$ are independent. It should be emphasized that the limit processes $\beta^{(1)}(t)$ and $\beta^{(2)}(t)$ are some functionals from the process ζ of Brownian motion in a bilayer environment with known explicit form of distributions. In particular, for $\alpha = 0$ and $\overline{b}(x) = b \operatorname{sign} x$ the limit process has the form $\beta^{(1)}(t) = bL_\zeta^0(t)$, where $L_\zeta^0(t)$ is the local time at zero of the process ζ in the interval $[0, t]$.

Note that similar limit processes of the form $\beta^{(1)}(t)$ and $\beta^{(2)}(t)$, where $\zeta = W$ is a Wiener process, were first obtained in the book [81].

In Theorems 4.3 and 4.4, which are, in some sense, analogues of the central limit theorem, the results about weak convergence, as $T \to +\infty$, of the processes $\beta_T^{(2)}(t)$ (Theorem 4.3) and $\beta_T^{(1)}(t)$ (Theorem 4.4) to a Wiener process are obtained. In Theorem 4.5 the sufficient conditions for the weak convergence, as $T \to +\infty$, of the processes $\beta_T^{(1)}(t)$ to the process $W^*\left(\widehat{\beta}^{(1)}(t)\right)$ are established, where the Wiener process $W^*(t)$ and the process $\widehat{\beta}^{(1)}(t)$ are independent.

In Sect. 4.2, using similar methods we study the asymptotic behavior, as $T \to +\infty$, of the functionals $\beta_T^{(1)}(t)$ and $\beta_T^{(2)}(t)$ of the solutions of Eq. (3.1). The obtained results are completely similar to the results of Sect. 4.1 with the only difference that the corresponding limit processes $\beta^{(1)}(t)$ and $\beta^{(2)}(t)$ are the functionals of the diffusion Bessel process $r(t)$. Note that in this case we know the explicit form of the distributions of the process $r(t)$.

In Sect. 4.3 the functionals of the form

$$I(t) = F(\xi(t)) + \int_0^t g(\xi(s)) \, dW(s)$$

are considered, where $F(x)$ is a continuous functions, $x \in \mathbb{R}$, $g(x)$ is a real valued and measurable function, which is locally square integrable, the processes ξ and W are related via Eq. (2.1) for $\sigma(x) \equiv 1$, $\xi(0) = x_0$ or via Eq. (3.1). Here the integral functional $\int_0^t g(\xi(s)) \, dW(s)$ is unbounded in probability. Sufficient conditions for the weak convergence, as $T \to +\infty$, of the processes $I_T(t) = \frac{I(tT)}{B(T)}$, $t > 0$ are formulated, where $B(T)$ is normalizing multiplier, $B(T) \to +\infty$, as $T \to +\infty$, to the functional

$$I_0(t) = \overline{a}(\zeta(t)) \, \zeta(t) \, |\zeta(t)|^\alpha + \int_0^t \overline{b}(\zeta(s)) \, |\zeta(s)|^\alpha \, d\zeta(s),$$

where ζ is the solution of Eq. (2.8) (Theorem 4.11) or to the functional

$$I_0^*(t) = \overline{a}(\zeta(t)) \, |\zeta(t)|^{\frac{1+\alpha_1}{2}} + W^*\left(\beta^{(1)}(t)\right),$$

where

$$\beta^{(1)}(t) = 2\left[\int_0^{\zeta(t)} \overline{b}(x) \, |x|^{\alpha_1} \, dx - \int_0^t \overline{b}(\zeta(s)) \, |\zeta(s)|^{\alpha_1} \, d\zeta(s)\right],$$

ζ is the solution of Eq. (2.8), W^* is a Wiener process, and the processes W^* and ζ are independent (Theorem 4.12).

The results of weak convergence, as $T \to +\infty$, of the processes I_T to the functionals I_0 and I_0^* of the diffusion Bessel process r are formulated in Theorems 4.13 and 4.14. If we replace ζ with r in functionals $I_0(t)$ and $I_0^*(t)$ from Theorems 4.11 and 4.12, we obtain the limit functionals $I_0(t)$ and $I_0^*(t)$ in Theorems 4.13 and 4.14, respectively.

Section 4.4 contains several examples that illustrate the results of previous sections.

4.1   Weak Convergence to the Functionals of a Brownian Motion in a Bilayer Environment

Now we study the behavior of the distributions, as $t \to +\infty$, of the integral functionals of the solutions ξ to Eq. (2.1), whose coefficients satisfy the conditions of Theorem 2.1.

In what follows we use the notation:

$$J(x) = \int\limits_0^x \frac{g(\varphi(u))}{[f'(\varphi(u)) \, \sigma(\varphi(u))]^2} \, du,$$

where

$$f'(x) = \exp\left\{ -2 \int\limits_0^x \frac{a(v)}{\sigma^2(v)} \, dv \right\},$$

$\varphi(x)$ is the inverse function to $f(x)$ from (2.1).

Definition 4.1 Let Ψ denote the class of functions $\psi(r) > 0$, $r \geq 0$, that are nondecreasing and regularly varying at infinity of order $\alpha \geq 0$, so that

$$\frac{\psi(rT)}{\psi(T)} \to r^\alpha,$$

as $T \to +\infty$ for all $r > 0$.

Theorem 4.1 *Let ξ be the solution to Eq. (2.1). Let the assumptions of Theorem 2.1 hold and*

$$\frac{1}{f(x)} \int\limits_0^x \frac{du}{f'(u)\sigma^2(u)} \to \begin{cases} \frac{1}{\sigma_1^2}, & x \to +\infty, \\ \frac{1}{\sigma_2^2}, & x \to -\infty. \end{cases}$$

Let the real-valued measurable function g be locally integrable and there exist a function $\psi = \psi(r) \in \Psi$ and constants b_i, $i = 1, 2$ for which

$$\lim_{|x| \to +\infty} \frac{1}{x} \int\limits_0^x q^2(v) \, dv = 0, \tag{4.3}$$

where

$$q(x) = \frac{1}{\psi(|x|)} J(x) - \overline{b}(x),$$

and

$$\bar{b}(x) = \begin{cases} b_1, & x \geq 0, \\ b_2, & x < 0. \end{cases}$$

Then the stochastic processes

$$\beta_T^{(1)}(t) = \frac{1}{\sqrt{T}\psi\left(\sqrt{T}\right)} \int_0^{tT} g\left(\xi(s)\right) ds$$

converge weakly, as $T \to +\infty$, to the process

$$\beta^{(1)}(t) = 2 \left[\int_0^{\zeta(t)} |u|^\alpha \bar{b}(u) \, du - \int_0^t |\zeta(s)|^\alpha \bar{b}(\zeta(s)) \, d\zeta(s) \right], \qquad (4.4)$$

where ζ is the solution to Eq. (2.8) and $\alpha \geq 0$ is the order of regularity of the function ψ at infinity.

Proof Consider the function

$$F(x) = 2 \int_0^{f(x)} J(u) \, du.$$

Note that the functions $F(x)$ and the derivative $F'(x)$ are continuous, the second derivative $F''(x)$ is locally integrable and

$$F'(x) a(x) + \frac{1}{2} F''(x) \sigma^2(x) = g(x)$$

a.e. with respect to the Lebesgue measure. Therefore, we can apply the Itô formula (Lemma A.3) to the process $F(\xi(tT))$ and obtain

$$\beta_T^{(1)}(t) = \frac{1}{\sqrt{T}\psi\left(\sqrt{T}\right)} \left[F\left(\xi(tT)\right) - F\left(\xi(0)\right) \right]$$

$$- \frac{1}{\sqrt{T}\psi\left(\sqrt{T}\right)} \int_0^{tT} F'\left(\xi(s)\right) \sigma\left(\xi(s)\right) dW(s).$$

Let us rewrite this equality in the form

$$\beta_T^{(1)}(t) = \frac{1}{\sqrt{T}\psi\left(\sqrt{T}\right)}\left[F\left(\varphi\left(\varsigma_T(t)\sqrt{T}\right)\right) - F\left(\varphi\left(\varsigma_T(0)\sqrt{T}\right)\right)\right]$$

$$-\frac{1}{\psi\left(\sqrt{T}\right)}\int_0^t F'\left(\varphi\left(\varsigma_T(s)\sqrt{T}\right)\right)\sigma\left(\varphi\left(\varsigma_T(s)\sqrt{T}\right)\right)dW_T(s).$$

According to the proof of Theorem 2.1, given an arbitrary sequence $T_n' \to +\infty$, we can choose a subsequence $T_n \to +\infty$, a probability space $(\widetilde{\Omega}, \widetilde{\mathfrak{F}}, \widetilde{P})$, and a stochastic process $\left(\widetilde{\varsigma}_{T_n}(t),\ \widetilde{W}_{T_n}(t)\right)$ defined on this space such that its finite-dimensional distributions coincide with those of the process $\left(\varsigma_{T_n}(t),\ W_{T_n}(t)\right)$ and, moreover,

$$\widetilde{\varsigma}_{T_n}(t) \xrightarrow{\widetilde{P}} \widetilde{\varsigma}(t), \quad \widetilde{W}_{T_n}(t) \xrightarrow{\widetilde{P}} \widetilde{W}(t),$$

as $T_n \to +\infty$, for all $t \geq 0$, where $\widetilde{\varsigma}(t)$, $\widetilde{W}(t)$ are some stochastic processes. Note that the processes $\left(\varsigma_{T_n}(t),\ W_{T_n}(t)\right)$ are weakly equivalent to the processes $\left(\widetilde{\varsigma}_{T_n}(t),\ \widetilde{W}_{T_n}(t)\right)$. Therefore, according to Corollary A.3, the processes $\beta_{T_n}^{(1)}(t)$ are weakly equivalent to the processes $\widetilde{\beta}_{T_n}^{(1)}(t)$, where

$$\widetilde{\beta}_{T_n}^{(1)}(t) = \frac{1}{\sqrt{T_n}\psi\left(\sqrt{T_n}\right)}\left[F\left(\varphi\left(\widetilde{\varsigma}_{T_n}(t)\sqrt{T_n}\right)\right) - F\left(\varphi\left(\widetilde{\varsigma}_{T_n}(0)\sqrt{T_n}\right)\right)\right]$$

$$-\frac{1}{\psi\left(\sqrt{T_n}\right)}\int_0^t F'\left(\varphi\left(\widetilde{\varsigma}_{T_n}(s)\sqrt{T_n}\right)\right)\sigma\left(\varphi\left(\widetilde{\varsigma}_{T_n}(s)\sqrt{T_n}\right)\right)d\widetilde{W}_{T_n}(s)$$

$$= 2\int_{\widetilde{\varsigma}_{T_n}(0)}^{\widetilde{\varsigma}_{T_n}(t)}\overline{b}(u)|u|^\alpha\,du - 2\int_0^t \overline{b}\left(\widetilde{\varsigma}_{T_n}(s)\right)\left|\widetilde{\varsigma}_{T_n}(s)\right|^\alpha d\widetilde{\varsigma}_{T_n}(s) + 2\sum_{k=1}^4 S_{T_n}^{(k)}(t), \qquad (4.5)$$

$$S_{T_n}^{(1)}(t) = \int_{\widetilde{\varsigma}_{T_n}(0)}^{\widetilde{\varsigma}_{T_n}(t)}\overline{b}(u)\left[\frac{\psi\left(|u|\sqrt{T_n}\right)}{\psi\left(\sqrt{T_n}\right)} - |u|^\alpha\right]du,$$

$$S_{T_n}^{(2)}(t) = \int_{\widetilde{\varsigma}_{T_n}(0)}^{\widetilde{\varsigma}_{T_n}(t)}q\left(u\sqrt{T_n}\right)\frac{\psi\left(|u|\sqrt{T_n}\right)}{\psi\left(\sqrt{T_n}\right)}du,$$

$$S_{T_n}^{(3)}(t) = -\int_0^t \bar{b}\left(\zeta_{T_n}(s)\right)\left[\frac{\psi\left(\left|\tilde{\zeta}_{T_n}(s)\right|\sqrt{T_n}\right)}{\psi\left(\sqrt{T_n}\right)} - \left|\tilde{\zeta}_{T_n}(s)\right|^\alpha\right]d\tilde{\zeta}_{T_n}(s),$$

and

$$S_{T_n}^{(4)}(t) = -\int_0^t q\left(\tilde{\zeta}_{T_n}(s)\sqrt{T_n}\right)\frac{\psi\left(\left|\tilde{\zeta}_{T_n}(s)\right|\sqrt{T_n}\right)}{\psi\left(\sqrt{T_n}\right)}d\tilde{\zeta}_{T_n}(s).$$

We have that $\tilde{\zeta}_{T_n}(t) \xrightarrow{\tilde{P}} \tilde{\zeta}(t)$, as $T_n \to +\infty$. Moreover, according to the proof of Theorem 2.1,

$$\lim_{N\to+\infty}\limsup_{T_n\to+\infty} P\left\{\sup_{0\le t\le L}\left|\tilde{\zeta}_{T_n}(t)\right| > N\right\} = 0,$$

$$\lim_{h\to 0}\limsup_{T_n\to+\infty} P\left\{\sup_{|t_1-t_2|\le h,\, t_i\le L}\left|\tilde{\zeta}_{T_n}(t_2) - \tilde{\zeta}_{T_n}(t_1)\right| > \varepsilon\right\} = 0$$

for arbitrary $\varepsilon > 0$, $L > 0$. The convergence $\tilde{\zeta}_{T_n}(t) \to \tilde{\zeta}(t)$, as $T_n \to +\infty$, is a weak convergence w.r.t. the measure $\tilde{P}$. Note that $\tilde{\zeta}_{T_n}(0) \to 0$, as $T_n \to +\infty$, with probability 1, and obtain

$$\int_{\tilde{\zeta}_{T_n}(0)}^{\tilde{\zeta}_{T_n}(t)} \bar{b}(u)\,|u|^\alpha\,du \xrightarrow{\tilde{P}} \int_0^{\tilde{\zeta}(t)} \bar{b}(u)\,|u|^\alpha\,du. \tag{4.6}$$

According to Lemma A.7,

$$\int_0^t \bar{b}\left(\tilde{\zeta}_{T_n}(s)\right)\left|\tilde{\zeta}_{T_n}(s)\right|^\alpha d\tilde{\zeta}_{T_n}(s) \xrightarrow{\tilde{P}} \int_0^t \bar{b}\left(\tilde{\zeta}(s)\right)\left|\tilde{\zeta}(s)\right|^\alpha d\tilde{\zeta}(s), \tag{4.7}$$

as $T_n \to +\infty$. Now, we are in a position to prove that

$$\sup_{0\le t\le L}\left|S_{T_n}^{(k)}(t)\right| \xrightarrow{\tilde{P}} 0, \quad k = \overline{1,4}. \tag{4.8}$$

Let

$$P_N = P\left\{\sup_{0\le t\le L}\left|\tilde{\zeta}_{T_n}(t)\right| > N\right\}.$$

In order to prove relations (4.8), we produce the following inequalities:

$$\mathsf{P}\left\{\sup_{0\leq t\leq L}\left|S_{T_n}^{(1)}(t)\right|>\varepsilon\right\}\leq P_N+\frac{2}{\varepsilon}\int_{-N}^{N}\left|\bar{b}(u)\right|\left|\frac{\psi\left(|u|\sqrt{T_n}\right)}{\psi\left(\sqrt{T_n}\right)}-|u|^\alpha\right|du, \quad (4.9)$$

$$\mathsf{P}\left\{\sup_{0\leq t\leq L}\left|S_{T_n}^{(2)}(t)\right|>\varepsilon\right\}\leq P_N+\frac{2}{\varepsilon}\int_{-N}^{N}q\left(u\sqrt{T_n}\right)\frac{\psi\left(|u|\sqrt{T_n}\right)}{\psi\left(\sqrt{T_n}\right)}du$$

$$(4.10)$$

$$\leq P_N+\frac{2}{\varepsilon}C_N\cdot\left(2N\frac{1}{\sqrt{T_n}}\int_{-N\sqrt{T_n}}^{N\sqrt{T_n}}q^2(u)\,du\right)^{\frac{1}{2}},$$

$$\mathsf{P}\left\{\sup_{0\leq t\leq L}\left|S_{T_n}^{(3)}(t)\right|>\varepsilon\right\}\leq P_N$$

$$+\left(\frac{2}{\varepsilon}\right)^2 4\mathsf{E}\int_0^L\left(\bar{b}\left(\tilde{\zeta}_{T_n}(s)\right)\left[\frac{\psi\left(\left|\tilde{\zeta}_{T_n}(s)\right|\sqrt{T_n}\right)}{\psi\left(\sqrt{T_n}\right)}-\left|\tilde{\zeta}_{T_n}(s)\right|^\alpha\right]\right)^2\hat{\sigma}_{T_n}^2\left(\tilde{\zeta}_{T_n}(s)\right)ds$$

$$\leq P_N+\left(\frac{2}{\varepsilon}\right)^2\hat{C}_N\int_0^L\mathsf{P}\left\{\left|\tilde{\zeta}_{T_n}(s)\right|<\delta\right\}ds$$

$$+\left(\frac{2}{\varepsilon}\right)^2 4C\mathsf{E}\int_0^L\left(\bar{b}\left(\tilde{\zeta}_{T_n}(s)\right)\left[\frac{\psi\left(\left|\tilde{\zeta}_{T_n}(s)\right|\sqrt{T_n}\right)}{\psi\left(\sqrt{T_n}\right)}-\left|\tilde{\zeta}_{T_n}(s)\right|^\alpha\right]\right)^2\chi_{0<\delta\leq\left|\tilde{\zeta}_{T_n}(s)\right|\leq N}\,ds,$$

$$(4.11)$$

$$\mathsf{P}\left\{\sup_{0\leq t\leq L}\left|S_{T_n}^{(4)}(t)\right|>\varepsilon\right\}\leq P_N$$

$$+\left(\frac{2}{\varepsilon}\right)^2 4\mathsf{E}\int_0^L q^2\left(\tilde{\zeta}_{T_n}(s)\sqrt{T_n}\right)\frac{\psi\left(\left|\tilde{\zeta}_{T_n}(s)\right|\sqrt{T_n}\right)}{\psi\left(\sqrt{T_n}\right)}\hat{\sigma}_{T_n}^2\left(\tilde{\zeta}_{T_n}(s)\right)\chi_{\left|\tilde{\zeta}_{T_n}(s)\right|\leq N}\,ds$$

$$\leq P_N+\left(\frac{2}{\varepsilon}\right)^2 4C_N\mathsf{E}\int_0^L q^2\left(\tilde{\zeta}_{T_n}(s)\sqrt{T_n}\right)\hat{\sigma}_{T_n}^2\left(\tilde{\zeta}_{T_n}(s)\right)\chi_{\left|\tilde{\zeta}_{T_n}(s)\right|\leq N}\,ds$$

$$(4.12)$$

for any $\varepsilon>0$, $L<+\infty$, $N<+\infty$.

The properties of regularly varying at infinity functions $\psi(r)$, $r\geq 0$ (Lemma A.17), Lemma 2.1 and the arbitrary choice of N, together with (4.9)–(4.11) imply relations (4.8) for $k=\overline{1,3}$.

Let us prove that

$$
\mathsf{E} \int_0^L q^2 \left(\tilde{\zeta}_{T_n} (s) \sqrt{T_n} \right) \hat{\sigma}_{T_n}^2 \left(\tilde{\zeta}_{T_n} (s) \right) \cdot \chi_{|\tilde{\zeta}_{T_n}(s)| \le N} \, ds \to 0, \tag{4.13}
$$

as $T_n \to +\infty$. Using the Itô formula we have

$$
\mathsf{E} \int_0^t q^2 \left(\tilde{\zeta}_{T_n} (s) \sqrt{T_n} \right) \hat{\sigma}_{T_n}^2 \left(\tilde{\zeta}_{T_n} (s) \right) \cdot \chi_{|\tilde{\zeta}_{T_n}(s)| \le N} \, ds
$$

$$
= \mathsf{E} \left[Q_{T_n} \left(\tilde{\zeta}_{T_n} (t) \right) - Q_{T_n} \left(\tilde{\zeta}_{T_n} (0) \right) \right],
$$

where

$$
Q_{T_n} (x) = 2 \int_0^x \left(\int_0^u q^2 \left(v \sqrt{T_n} \right) \chi_{|v| \le N} \, dv \right) du.
$$

Since

$$
\left| Q_{T_n}(x) \right| \le \frac{2}{\sqrt{T_n}} \int_{-N\sqrt{T_n}}^{N\sqrt{T_n}} q^2(v) \, dv \, |x|,
$$

taking into account the conditions of the theorem we have that

$$
\lim_{T_n \to +\infty} \mathsf{E} \left[Q_{T_n} \left(\tilde{\zeta}_{T_n} (t) \right) - Q_{T_n} \left(\tilde{\zeta}_{T_n} (0) \right) \right] = 0
$$

and, consequently, the relation (4.13). So, taking into account (4.12) we get (4.8) for $k = 4$.

Using relations (4.6)–(4.8) we conclude that (4.5) implies the following convergence:

$$
\tilde{\beta}_{T_n}^{(1)} (t) \xrightarrow{\tilde{\mathsf{P}}} \tilde{\beta}^{(1)} (t),
$$

as $T_n \to +\infty$, where

$$
\tilde{\beta}^{(1)} (t) = 2 \left[\int_0^{\tilde{\zeta}(t)} \bar{b} (u) \, |u|^\alpha \, du - \int_0^t \left| \tilde{\zeta} (s) \right|^\alpha \bar{b} \left(\tilde{\zeta} (s) \right) d\tilde{\zeta} (s) \right], \tag{4.14}
$$

and $\tilde{\zeta} (t)$ is the solution of Eq. (2.19).

Thus, we obtain the convergence of the distribution of the processes $\beta_{T_n}^{(1)}(t)$, as $T_n \to +\infty$, to the corresponding distribution of the process $\beta^{(1)}(t)$. Since the sequence $T_n \to +\infty$ is arbitrary and since the solution to Eq. (2.19) is unique, then we have the weak convergence of the finite-dimensional distributions of the process $\beta_T^{(1)}(t)$, as $T \to +\infty$, to the corresponding distribution of the process $\beta^{(1)}(t)$. Next, let us show that for any $L > 0$

$$\lim_{h \to 0} \limsup_{T \to +\infty} \mathsf{P} \left\{ \sup_{|t_1 - t_2| \le h, \, t_i \le L} \left| \beta_T^{(1)}(t_2) - \beta_T^{(1)}(t_1) \right| > \varepsilon \right\} = 0. \qquad (4.15)$$

It is clear that the relation (4.15) holds for $S_{T_n}^{(k)}(t)$, $k = \overline{1, 4}$ in the representation (4.5), as $T_n \to +\infty$. Using the inequality

$$\mathsf{P} \left\{ \sup_{|t_1 - t_2| \le h, \, t_i \le L} \left| \int_{\tilde{\zeta}_{T_n}(t_1)}^{\tilde{\zeta}_{T_n}(t_2)} \overline{b}(u) \, |u|^\alpha \, du \right| > \varepsilon \right\} \le P_N$$

$$+ \mathsf{P} \left\{ C_N \sup_{|t_1 - t_2| \le h, \, t_i \le L} \left| \tilde{\zeta}_{T_n}(t_2) - \tilde{\zeta}_{T_n}(t_1) \right| > \frac{\varepsilon}{2} \right\}$$

and the convergence (2.20) we have that the relation (4.15) holds for the process

$$\int_{\tilde{\zeta}_{T_n}(0)}^{\tilde{\zeta}_{T_n}(t)} \overline{b}(u) \, |u|^\alpha \, du.$$

Next, similarly to the proof of relation (2.20), we obtain

$$\mathsf{P} \left\{ \sup_{|t_1 - t_2| \le h, \, t_i \le L} \left| \int_{t_1}^{t_2} \overline{b}\left(\tilde{\zeta}_{T_n}(s) \right) \left| \tilde{\zeta}_{T_n}(s) \right|^\alpha d\tilde{\zeta}_{T_n}(s) \right| > \varepsilon \right\} \le P_N$$

$$\mathsf{P} \left\{ \sup_{|t_1 - t_2| \le h, \, t_i \le L} \left| \int_{t_1}^{t_2} \overline{b}\left(\tilde{\zeta}_{T_n}(s) \right) \left| \tilde{\zeta}_{T_n}(s) \right|^\alpha \chi_{|\tilde{\zeta}_{T_n}(s)| \le N} \, d\tilde{\zeta}_{T_n}(s) \right| > \frac{\varepsilon}{2} \right\} \le P_N$$

$$+ \sum_{kh \le L} \left(\frac{8}{\varepsilon} \right)^4 \mathsf{E} \left[\int_{kh}^{(k+1)h} \overline{b}\left(\tilde{\zeta}_{T_n}(s) \right) \left| \tilde{\zeta}_{T_n}(s) \right|^\alpha \chi_{|\tilde{\zeta}_{T_n}(s)| \le N} \, \widehat{\sigma}_{T_n}\left(\tilde{\zeta}_{T_n}(s) \right) d\widetilde{W}_{T_n}(s) \right]^4$$

$$\le P_N + C_N \sum_{kh \le L} h^2.$$

Consequently, the relation (4.15) holds for all terms of the right-hand side of equality (4.5). Therefore, the relation (4.15) holds for the process $\beta_{T_n}^{(1)}(t)$. Since the sequence T_n' is arbitrary, the relation (4.15) holds for the process $\beta_T^{(1)}(t)$, as well. According to Theorem A.13, the stochastic processes $\beta_T^{(1)}$ converge weakly, as $T \to +\infty$, to the process $\beta^{(1)}$. □

We emphasize that for $\alpha = 0$ and $\bar{b}(x) = b \operatorname{sign} x$ the limit process has the form $\beta^{(1)}(t) = bL_\zeta^0(t)$, where $L_\zeta^0(t)$ is the local time at zero of the process ζ on the interval $[0, t]$. For $\alpha = 1$ and $\bar{b}(x) = b \operatorname{sign} x$ the limit process has the form $\beta^{(1)}(t) = b \int_0^t \bar{\sigma}^2\left(\widetilde{\zeta}(s)\right) ds$, where $\bar{\sigma}(x)$ is defined in (2.8). In particular, for $\bar{\sigma}^2(x) = \sigma_0^2$ the limit process is $\beta^{(1)}(t) = b\sigma_0^2 t$.

Theorem 4.2 *Let (ξ, W) be a solution of Eq. (2.1). Let the assumptions of Theorem 2.1 hold and*

$$\frac{1}{f(x)} \int_0^x \frac{du}{f'(u)\sigma^2(u)} \to \begin{cases} \frac{1}{\sigma_1^2}, & x \to +\infty, \\ \frac{1}{\sigma_2^2}, & x \to -\infty. \end{cases}$$

Let the real-valued measurable function $g = g(x)$ be locally square integrable and such that there exist functions $\psi_i \in \Psi$ of orders $\alpha_i \geq 0$, respectively, and constants b_i, $i = 1, 2$, and $\mathsf{E}\,|f(\xi(0))|^{\delta+\alpha_2} < +\infty$, for some $\delta > 1$,

$$\lim_{|x| \to +\infty} \frac{1}{x} \int_0^x q^2(v)\, dv = 0, \tag{4.16}$$

where

$$q(x) = \frac{1}{\psi_1(|x|)} \int_0^x \frac{g^2(\varphi(u))}{[f'(\varphi(u))\,\sigma(\varphi(u))]^2} du - \bar{b}(x)$$

with

$$\bar{b}(x) = \begin{cases} b_1, & x \geq 0, \\ b_2, & x < 0, \end{cases}$$

and

$$\left| \frac{1}{\psi_2(|x|)} \int_0^x \frac{|g(\varphi(v))|}{[f'(\varphi(v))\,\sigma(\varphi(v))]^2} dv \right| \leq C, \tag{4.17}$$

$$\lim_{T \to +\infty} \frac{\psi_2 \left(\sqrt{T} \right)}{T^{\frac{1}{4}} \left(\psi_1 \left(\sqrt{T} \right) \right)^{\frac{1}{2}}} = 0. \tag{4.18}$$

Then the stochastic processes

$$\beta_T^{(2)} (t) = \frac{1}{T^{\frac{1}{4}} \left(\psi_1 \left(\sqrt{T} \right) \right)^{\frac{1}{2}}} \int_0^{tT} g \left(\xi (s) \right) \, dW (s)$$

converge weakly, as $T \to +\infty$, *to the process* $\beta^{(2)} (t) = W^* \left(\beta^{(1)}(t) \right)$, *where the process* $\beta^{(1)}$ *has the form (4.4) for* $\alpha = \alpha_1$, W^* *is a Wiener process, and the processes* W^* *and* $\beta^{(1)}$ *are independent.*

Proof Let $\tilde{\zeta}_{T_n} (t)$ and $\tilde{W}_{T_n} (t)$ be the processes introduced in the proof of Theorem 4.1. According to Corollary A.3, the processes $\beta_{T_n}^{(2)} (t)$ are weakly equivalent to the processes

$$\tilde{\beta}_{T_n}^{(2)} (t) = \frac{\sqrt{T_n}}{T_n^{\frac{1}{4}} \left(\psi_1 \left(\sqrt{T_n} \right) \right)^{\frac{1}{2}}} \int_0^t g \left(\varphi \left(\tilde{\zeta}_{T_n} (s) \sqrt{T_n} \right) \right) \, d\tilde{W}_{T_n} (s)$$

$$= \frac{\sqrt[4]{T_n}}{\left(\psi_1 \left(\sqrt{T_n} \right) \right)^{\frac{1}{2}}} \int_0^t g \left(\varphi \left(\tilde{\zeta}_{T_n} (s) \sqrt{T_n} \right) \right) \, d\tilde{W}_{T_n} (s).$$

Let $\tau_{T_n} (t)$ be the minimal solution to the equation

$$\tilde{\beta}_{T_n}^{(1)} \left(\tau_{T_n} (t) \right) = t,$$

where

$$\tilde{\beta}_{T_n}^{(1)} (t) = \frac{\sqrt{T_n}}{\psi_1 (\sqrt{T_n})} \int_0^t g^2 \left(\varphi \left(\tilde{\zeta}_{T_n} (s) \sqrt{T_n} \right) \right) \, ds.$$

Note that for every T_n and t, the random variable $\tau_{T_n} (t)$ is a Markov moment with respect to the σ-algebras $\sigma \left\{ \tilde{\zeta}_{T_n} (s), \ s \leq t \right\}$. According to Lemma A.8,

$$\int_0^{+\infty} g^2 \left(\varphi \left(\tilde{\zeta}_{T_n} (s) \sqrt{T_n} \right) \right) \, ds = +\infty$$

with probability 1. Thus, given an arbitrary T_n, the process $\widetilde{\beta}^{(1)}_{T_n}(t)$ attains any positive level with probability 1. Hence $\widetilde{\beta}^{(2)}_{T_n}\left(\tau_{T_n}(t)\right)$ is a sequence of Wiener processes (see Theorem A.8). Denote these processes by $W^*_{T_n}(t)$.

Now, $\widetilde{\beta}^{(2)}_{T_n}(t) = W^*_{T_n}\left(\tau^{-1}_{T_n}(t)\right)$, where $\tau^{-1}_{T_n}(t) = \widetilde{\beta}^{(1)}_{T_n}(t)$, that is,

$$\widetilde{\beta}^{(2)}_{T_n}(t) = W^*_{T_n}\left(\widetilde{\beta}^{(1)}_{T_n}(t)\right).$$

It follows from the proof of Theorem 4.1 that $\widetilde{\beta}^{(1)}_{T_n}(t) \xrightarrow{\widetilde{P}} \widetilde{\beta}^{(1)}(t)$, as $T_n \to +\infty$, where $\widetilde{\beta}^{(1)}(t)$ is given by (4.14) for $\alpha = \alpha_1$. The processes $\widetilde{\beta}^{(1)}(t)$ are measurable with respect to the σ-algebra $\sigma\left\{\widetilde{\zeta}(s),\ s \leq t\right\}$.

Using the inequality

$$\widetilde{P}\left\{\left|W^*_{T_n}\left(\widetilde{\beta}^{(1)}_{T_n}(t)\right) - W^*_{T_n}\left(\widetilde{\beta}^{(1)}(t)\right)\right| > \varepsilon\right\}$$

$$\leq \widetilde{P}\left\{\widetilde{\beta}^{(1)}_{T_n}(t) > N\right\} + \widetilde{P}\left\{\beta^{(1)}(t) > N\right\}$$

$$+ \widetilde{P}\left\{\left|\widetilde{\beta}^{(1)}_{T_n}(t) - \widetilde{\beta}^{(1)}(t)\right| > \delta\right\} + \widetilde{P}\left\{\sup_{|t_1-t_2|\leq\delta,\, t_i\leq N}\left|W^*_{T_n}(t_2) - W^*_{T_n}(t_1)\right| > \varepsilon\right\}$$

and the convergence

$$\lim_{\delta\to 0}\sup_{T_n}\widetilde{P}\left\{\sup_{|t_1-t_2|\leq\delta,\, t_i\leq N}\left|W^*_{T_n}(t_2) - W^*_{T_n}(t_1)\right| > \varepsilon\right\} = 0$$

for any $\varepsilon > 0$, $N > 0$ (see [19, Chapter IX, § 3]), we obtain that

$$W^*_{T_n}\left(\widetilde{\beta}^{(1)}_{T_n}(t)\right) - W^*_{T_n}\left(\widetilde{\beta}^{(1)}(t)\right) \xrightarrow{\widetilde{P}} 0,$$

as $T_n \to +\infty$.

Therefore,

$$\widetilde{\beta}^{(2)}_{T_n}(t) - W^*_{T_n}\left(\widetilde{\beta}^{(1)}(t)\right) \xrightarrow{\widetilde{P}} 0, \tag{4.19}$$

as $T_n \to +\infty$.

The proof of Theorem 2.1 and the representation (2.19) imply that the process

$$\eta_{T_n}(t) = \int_0^t \frac{d\widetilde{\zeta}_{T_n}(s)}{\overline{\sigma}\left(\widetilde{\zeta}_{T_n}(s)\right)}$$

converges, as $T_n \to +\infty$, in probability to a Wiener process $\widehat{W}(t)$ for every $t > 0$.

Using properties of stochastic integrals (see Theorem A.6), together with the fact that $\tau_{T_n}(t)$ is a Markov moment, and with the inequality

$$\left| \frac{\widehat{\sigma}_{T_n}(x)}{\overline{\sigma}(x)} \right| \leq C,$$

we get

$$\left| \mathsf{E} \eta_{T_n}(t) W_{T_n}^*(t) \right|$$

$$= \left| \mathsf{E} \int_0^t \frac{\widehat{\sigma}_{T_n}\left(\widetilde{\zeta}_{T_n}(s)\right)}{\overline{\sigma}\left(\widetilde{\zeta}_{T_n}(s)\right)} d\widetilde{W}_{T_n}(s) \frac{\sqrt{T_n}}{T_n^{\frac{1}{4}} \left(\psi_1\left(\sqrt{T_n}\right)\right)^{\frac{1}{2}}} \int_0^{\tau_{T_n}(t)} g\left(\varphi\left(\widetilde{\zeta}_{T_n}(s)\sqrt{T_n}\right)\right) d\widetilde{W}_{T_n}(s) \right|$$

$$= \frac{\sqrt[4]{T_n}}{\left(\psi_1\left(\sqrt{T_n}\right)\right)^{\frac{1}{2}}} \left| \mathsf{E} \int_0^{t \wedge \tau_{T_n}(t)} \frac{\widehat{\sigma}_{T_n}\left(\widetilde{\zeta}_{T_n}(s)\right)}{\overline{\sigma}\left(\widetilde{\zeta}_{T_n}(s)\right)} g\left(\varphi\left(\widetilde{\zeta}_{T_n}(s)\sqrt{T_n}\right)\right) ds \right|$$

$$\leq \frac{C\psi_2(\sqrt{T_n})}{T_n^{\frac{1}{4}} \left(\psi_1\left(\sqrt{T_n}\right)\right)^{\frac{1}{2}}} \mathsf{E} \frac{\sqrt{T_n}}{\psi_2(\sqrt{T_n})} \int_0^t \left| g\left(\varphi\left(\widetilde{\zeta}_{T_n}(s)\sqrt{T_n}\right)\right) \right| ds.$$

$$(4.20)$$

For the functional under the sign of the expectation in the last term of (4.20), we can write a representation similar to representation (4.5). Using such a representation and taking into account the properties of regular at infinity functions, together with the conditions of Theorem 4.2, we obtain that for sufficiently large T_n

$$\mathsf{E} \frac{\sqrt{T_n}}{\psi_2(\sqrt{T_n})} \int_0^t \left| g\left(\varphi\left(\widetilde{\zeta}_{T_n}(s)\sqrt{T_n}\right)\right) \right| ds$$

$$\leq C_N \mathsf{E} \left| \widetilde{\zeta}_{T_n}(t) - \widetilde{\zeta}_{T_n}(0) \right| + C_{\alpha_2} \mathsf{E} \left| \left| \widetilde{\zeta}_{T_n}(t) \right|^{\delta+\alpha_2} + \left| \widetilde{\zeta}_{T_n}(0) \right|^{\delta+\alpha_2} \right| \leq \widehat{C}_1 + \widehat{C}_2 t^{\frac{\delta+\alpha_2}{2}}$$

for certain constants $\widehat{C}_i$, $i = 1, 2$. Consequently,

$$\lim_{T_n \to +\infty} \mathsf{E} \eta_{T_n}(t) W_{T_n}^*(t) = \lim_{T_n \to +\infty} \mathsf{E} \widehat{W}(t) W_{T_n}^*(t) = 0. \qquad (4.21)$$

Since $\widehat{W}$ and $W_{T_n}^*$ are Wiener processes and they are asymptotically uncorrelated at any point, so, due to independency of their increments, they are asymptotically uncorrelated at any pair of points. So, their mutual quadratic characteristic is zero. Therefore, according to the Levy theorem (see [20, Chapter 1, § 3, Theorem 3]) we conclude that, as $T_n \to +\infty$, $W_{T_n}^*$ does not depend asymptotically on $\widehat{W}$. Further, the process $\widetilde{\beta}^{(1)}$ generates the same filtration as the process $\widehat{W}$, whence we conclude that $W^*_{T_n}$, as $T_n \to +\infty$, does not depend asymptotically on $\widetilde{\beta}^{(1)}$. It is easy to show that, as $T_n \to +\infty$, the finite-dimensional distributions of the process $W^*_{T_n}\left(\widetilde{\beta}^{(1)}\right)$ coincide with the corresponding finite-dimensional distributions of the process $W^*\left(\widetilde{\beta}^{(1)}\right)$, where W^* is a Wiener process, and the processes W^* and $\widetilde{\beta}^{(1)}$ are independent.

Therefore, according to the convergence (4.19), the finite-dimensional distributions of the process $\beta^{(2)}_{T_n}$ converge, as $T_n \to +\infty$, to the corresponding finite-dimensional distributions of the process $W^*\left(\widetilde{\beta}^{(1)}\right)$.

Furthermore, for every $L > 0$ we have the inequality

$$
\mathsf{P}\left\{ \sup_{|t_1-t_2|\leq h,\, t_i \leq L} \left| \widetilde{\beta}^{(2)}_{T_n}(t_2) - \widetilde{\beta}^{(2)}_{T_n}(t_1) \right| > \varepsilon \right\}
$$

$$
\leq \mathsf{P}\left\{ \sup_{|t_1-t_2|\leq h,\, t_i \leq L} \left| W^*_{T_n}\left(\widetilde{\beta}^{(1)}_{T_n}(t_2)\right) - W^*_{T_n}\left(\widetilde{\beta}^{(1)}_{T_n}(t_1)\right) \right| > \varepsilon \right\}
$$

$$
\leq \mathsf{P}\left\{ \sup_{0\leq t\leq L} \widetilde{\beta}^{(1)}_{T_n}(t) > N \right\} + \mathsf{P}\left\{ \sup_{|t_1-t_2|\leq h,\, t_i \leq L} \left| \widetilde{\beta}^{(1)}_{T_n}(t_2) - \widetilde{\beta}^{(1)}_{T_n}(t_1) \right| > \delta \right\}
$$

$$
+ \mathsf{P}\left\{ \sup_{|t_1-t_2|\leq \delta,\, 0\leq t_i \leq N} \left| W^*_{T_n}(t_2) - W^*_{T_n}(t_1) \right| > \varepsilon \right\}
$$

for any $\varepsilon > 0$, $N > 0$.

From this inequality and the convergence (4.15) we have for the process $\widetilde{\beta}^{(2)}_{T_n}$ a convergence analogous to convergence (4.15). Thus, according to Theorem A.13, the process $\widetilde{\beta}^{(2)}_{T_n}$ converges weakly, as $T_n \to +\infty$, to the process $W^*\left(\widetilde{\beta}^{(1)}(\cdot)\right)$. Since the subsequence $T_n \to +\infty$ is arbitrary and since the solution $\widetilde{\zeta}$ to Eq. (2.19) is unique, the proof of Theorem 4.2 is complete. $\qquad\square$

Theorem 4.3 *Let (ξ, W) be a solution of Eq. (2.1). Let the assumptions of Theorem 2.1 hold and*

$$
\frac{1}{f(x)} \int_0^x \frac{du}{f'(u)\sigma^2(u)} \to \begin{cases} \dfrac{1}{\sigma_1^2}, & x \to +\infty, \\[2mm] \dfrac{1}{\sigma_2^2}, & x \to -\infty. \end{cases}
$$

Let the real-valued measurable function $g = g(x)$ be locally square integrable and such that there exist constants b_i, $i = 1, 2$, for which

$$
\lim_{|x|\to +\infty} \frac{1}{x} \int_0^x q^2(v)\, dv = 0,
$$

where

$$q(x) = \frac{1}{|x|} \int_0^x \frac{g^2(\varphi(u))}{[f'(\varphi(u)) \sigma(\varphi(u))]^2} du - \overline{b}(x).$$

Here

$$\overline{b}(x) = \begin{cases} b_1, & x \geq 0, \\ b_2, & x < 0 \end{cases}$$

and satisfy the following relation:

$$|\overline{b}(x)| \overline{\sigma}^2(x) = \sigma_0^2$$

for all $x \in \mathbb{R}$ and for some constant σ_0. Then the stochastic processes

$$\beta_T^{(2)}(t) = \frac{1}{\sqrt{T}} \int_0^{tT} g(\xi(s)) \, dW(s)$$

converge weakly, as $T \to +\infty$, to the process $\sigma_0 W^*(t)$, where W^* is a Wiener process.

Proof It is clear that the function g satisfies condition (4.16) of Theorem 4.2 for $\psi_1(r) \in \Psi$ with $\alpha_1 = 1$. Note that we do not need now the conditions (4.17) and (4.18) because they used in the proof of Theorem 4.2 only to prove the independence of the processes $W_{T_n}^*(t)$ and $\widetilde{\beta}^{(1)}(t)$. But now the limit process $\widetilde{\beta}^{(1)}(t)$ is degenerate and it is not necessary to establish any independence.

According to the Itô formula, in the case $\alpha = 1$ and $|\overline{b}(x)| \overline{\sigma}^2(x) = \sigma_0^2$, we obtain that $\widetilde{\beta}^{(1)}(t) = \sigma_0^2 t$. Using relation (4.19) and the fact that the finite-dimensional distributions of the process $W_{T_n}^*(\sigma_0^2 t)$ coincide with the corresponding finite-dimensional distributions of the process $W^*(\sigma_0^2 t)$, where $W^*(t)$ is a Wiener process, we have the convergence of the finite-dimensional distributions of the process $\beta_{T_n}^{(2)}(t)$ to the corresponding finite-dimensional distributions of the process $\sigma_0 W^*(t)$.

Now we conclude about the weak convergence for the same reasons as those in the proof of Theorem 4.2. □

Note that Theorem 4.1 does not hold for a functional of the form $\int_0^t \sin W(s) ds$. For the functions $g(x)$ with oscillatory properties the following theorem holds.

Theorem 4.4 *Let ξ be the solution to Eq. (2.1). Let the assumptions of Theorem 2.1 hold and*

$$\frac{1}{f(x)} \int\limits_0^x \frac{du}{f'(u)\sigma^2(u)} \to \begin{cases} \frac{1}{\sigma_1^2}, & x \to +\infty, \\ \frac{1}{\sigma_2^2}, & x \to -\infty. \end{cases}$$

Let the real-valued measurable function $g(x)$ be locally bounded and such that there exist some constant $C > 0$ that bounds $J(x)$ for all x:

$$|J(x)| \le C.$$

Assume that there exist constants $\sigma_0 > 0$ and $c_0 \in \mathbb{R}$ such that

$$\lim_{|x| \to +\infty} \frac{1}{|x|} \int\limits_0^x (J(u) - c_0)\, du = 0,$$

and

$$\frac{1}{|x|} \int\limits_0^x [J(u) - c_0]^2\, du \to \begin{cases} \dfrac{\sigma_0^2}{\sigma_1^2}, & x \to +\infty, \\ -\dfrac{\sigma_0^2}{\sigma_2^2}, & x \to -\infty. \end{cases}$$

Then the stochastic processes

$$\beta_T^{(1)}(t) = \frac{1}{\sqrt{T}} \int\limits_0^{tT} g\,(\xi(s))\, ds$$

converge weakly, as $T \to +\infty$, to the process $\beta^{(1)}(t) = 2\sigma_0 W^*(t)$, where W^* is a Wiener process.

Proof Using the Itô formula, we obtain

$$\beta_T^{(1)}(t) = \frac{1}{\sqrt{T}} [F\,(\xi(tT)) - F\,(\xi(0))] - \frac{1}{\sqrt{T}} \int\limits_0^{tT} F'\,(\xi(s))\, \sigma\,(\xi(s))\, dW(s),$$

where

$$F(x) = 2 \int\limits_0^{f(x)} (J(u) - c_0)\, du.$$

Note that $f^{-1}(x)F(x) \to 0$, as $|f(x)| \to +\infty$. Therefore, for every $\varepsilon > 0$, there exists $N_\varepsilon > 0$ such that for all $|f(x)| > N_\varepsilon$ one has $\left|f^{-1}(x)F(x)\right| < \varepsilon$.

Using the inequalities

$$\sup_{0 \le t \le L} \left| \frac{F\left(\xi(tT)\right)}{\sqrt{T}} \right| \left[\chi_{|f(\xi(tT))| \le N_\varepsilon} + \chi_{|f(\xi(tT))| > N_\varepsilon} \right]$$

$$\le \sup_{0 \le t \le L} \left| \frac{F\left(\xi(tT)\right)}{\sqrt{T}} \right| \chi_{|f(\xi(tT))| > N_\varepsilon} + \sup_{0 \le t \le L} \left| \frac{F\left(\xi(tT)\right)}{\sqrt{T}} \right| \chi_{|f(\xi(tT))| \le N_\varepsilon}$$

$$\le \varepsilon \sup_{0 \le t \le L} |\zeta_T(t)| + \sup_{|x| \le N_\varepsilon} |F(\varphi(x))| \frac{1}{\sqrt{T}} \le \varepsilon \sup_{0 \le t \le L} |\zeta_T(t)| + C_\varepsilon \frac{1}{\sqrt{T}},$$

we obtain

$$\limsup_{T \to +\infty} \mathsf{E} \sup_{0 \le t \le L} \left| \frac{F\left(\xi(tT)\right)}{\sqrt{T}} \right| \le C_L \varepsilon$$

for any $\varepsilon > 0$.

Consequently,

$$\mathsf{E} \sup_{0 \le t \le L} \left| \frac{F\left(\xi(tT)\right)}{\sqrt{T}} \right| \to 0,$$

as $T \to +\infty$. For arbitrary $L > 0$ we have

$$\sup_{0 \le t \le L} \left| \beta_T^{(1)}(t) + \frac{1}{\sqrt{T}} \int_0^{tT} F'\left(\xi(s)\right) \sigma\left(\xi(s)\right) dW(s) \right| \overset{\mathsf{P}}{\longrightarrow} 0, \qquad (4.22)$$

as $T \to +\infty$.

The conditions of Theorem 4.3 hold for the function $\left(F'(x)\sigma(x)\right)^2$ with $\left|\bar{b}(x)\right| \bar{\sigma}^2(x) = 4\sigma_0^2$. That is why the stochastic process

$$\frac{1}{\sqrt{T}} \int_0^{tT} F'\left(\xi(s)\right) \sigma\left(\xi(s)\right) dW(s)$$

converges weakly, as $T \to +\infty$, to the process $2\sigma_0 W^*(t)$, where $W^*(t)$ is a Wiener process.

Taking into account (4.22), we obtain the proof of Theorem 4.4.   □

Theorem 4.5 *Let ξ be a solution to Eq. (2.1) with initial condition $\xi(0) = x_0$. Let the assumptions of Theorem 2.1 hold and*

$$\frac{1}{f(x)} \int_0^x \frac{du}{f'(u)\sigma^2(u)} \to \begin{cases} \frac{1}{\sigma_1^2} , & x \to +\infty, \\ \frac{1}{\sigma_2^2} , & x \to -\infty. \end{cases}$$

Let the real-valued measurable function $g(x)$ be locally bounded and such that there exist constants $\hat{b}_i$, $i = 1, 2$ and $c_0 \in \mathbb{R}$, together with functions $\psi_i(r) \in \Psi$, $i = 1, 2$ of order $\alpha_i \geq 0$, respectively, for which

(i_1)

$$\lim_{|x| \to +\infty} \frac{1}{\sqrt{|x|\, \psi_1\,(|x|)}} \int_0^x [J(u) - c_0]\, du = 0,$$

and
(i_2)

$$\lim_{|x| \to +\infty} \left[\frac{1}{\psi_1\,(|x|)} \int_0^x [J(u) - c_0]^2\, du - \hat{b}(x) \right] = 0,$$

where

$$\hat{b}(x) = \begin{cases} \hat{b}_1, & x \geq 0, \\ \hat{b}_2, & x < 0. \end{cases}$$

(i_3) *Also, let for some constant $C > 0$ the following inequality hold for all $x \in \mathbb{R}$:*

$$\frac{1}{\psi_2\,(|x|)} \left| \int_0^x \frac{1}{f'\,(\varphi\,(u))\,\sigma\,(\varphi\,(u))} |J(u) - c_0|\, du \right| \leq C,$$

and
(i_4)

$$\lim_{T \to +\infty} \frac{\psi_2\left(\sqrt{T}\right)}{T^{\frac{1}{4}}\left(\psi_1\left(\sqrt{T}\right)\right)^{\frac{1}{2}}} = 0.$$

Then the stochastic processes

$$\beta_T^{(1)}(t) = \frac{1}{T^{\frac{1}{4}}\left(\psi_1\left(\sqrt{T}\right)\right)^{\frac{1}{2}}} \int_0^{tT} g\left(\xi(s)\right) ds$$

converge weakly, as $T \to +\infty$*, to the process* $W^*\left(\widehat{\beta}^{(1)}(t)\right)$*, where* W^* *is a Wiener process,*

$$\widehat{\beta}^{(1)}(t) = 8\left[\int_0^{\zeta(t)} |u|^{\alpha_1} \hat{b}(u)\, du - \int_0^t |\zeta(s)|^{\alpha_1} \hat{b}(\zeta(s))\, d\zeta(s)\right], \qquad (4.23)$$

and ζ *is the solution to Eq.(2.8), the Wiener process* W^* *and the process* $\widehat{\beta}^{(1)}$ *are independent.*

Proof Using the Itô formula, we obtain

$$\beta_T^{(1)}(t) = \frac{1}{T^{\frac{1}{4}}\left(\psi_1\left(\sqrt{T}\right)\right)^{\frac{1}{2}}} [F\left(\xi(tT)\right) - F\left(\xi(0)\right)]$$

$$-\frac{1}{T^{\frac{1}{4}}\left(\psi_1\left(\sqrt{T}\right)\right)^{\frac{1}{2}}} \int_0^{tT} F'\left(\xi(s)\right) \sigma\left(\xi(s)\right) dW(s),$$

where

$$F(x) = 2\int_0^{f(x)} (J(u) - c_0)\, du$$

and

$$\lim_{|x| \to +\infty} \frac{1}{\sqrt{|f(x)|\,\psi_1\left(|f(x)|\right)}} F(x) = 0.$$

Consequently,

$$\sup_{0 \le t \le L} \frac{1}{T^{\frac{1}{4}}\left(\psi_1\left(\sqrt{T}\right)\right)^{\frac{1}{2}}} |F\left(\xi(tT)\right)| \xrightarrow{P} 0,$$

as $T \to +\infty$.

Using the condition (i_1), similarly to (4.22) we get for arbitrary $L > 0$

$$\sup_{0 \le t \le L} \left| \beta_T^{(1)}(t) + \beta_T^{(2)}(t) \right| \xrightarrow{P} 0, \tag{4.24}$$

as $T \to +\infty$, where

$$\beta_T^{(2)}(t) = \frac{1}{T^{\frac{1}{4}} \left(\psi_1 \left(\sqrt{T} \right) \right)^{\frac{1}{2}}} \int_0^{tT} F'(\xi(s)) \sigma(\xi(s)) \, dW(s).$$

According to conditions (i_2)–(i_4), it is easy to obtain that the function $[F'(x)\sigma(x)]$ satisfy the assumptions of Theorem 4.2 for $\overline{b}(x) = 4\hat{b}(x)$. Indeed, taking into account the equalities $F'(x) = 2[J(f(x)) - c_0] f'(x)$ and $F'(\varphi(x)) = 2[J(x) - c_0] f'(\varphi(x))$ we have that the function

$$q(x) = \frac{1}{\psi_1(|x|)} \int_0^x \frac{\left[F'(\varphi(u)) \sigma(\varphi(u)) \right]^2}{[f'(\varphi(u)) \sigma(\varphi(u))]^2} du - \overline{b}(x)$$

from Theorem 4.2 has the form

$$q(x) = \frac{4}{\psi_1(|x|)} \int_0^x [J(u) - c_0]^2 \, du - \overline{b}(x).$$

Applying the condition (i_2), we have that $q(x) \to 0$, as $|x| \to +\infty$, and consequently $\frac{1}{x} \int_0^x q^2(u) \, du \to 0$, as $|x| \to +\infty$. Therefore, the function $[F'(x)\sigma(x)]^2$ satisfies the assumptions of Theorem 4.1 for $\overline{b}(x) = 4\hat{b}(x)$. Taking into account the assumption (i_3) we obtain that

$$\frac{1}{\psi_2(|x|)} \left| \int_0^x \frac{|F'(\varphi(u)) \sigma(\varphi(u))|}{[f'(\varphi(u)) \sigma(\varphi(u))]^2} du \right|$$

$$= \frac{2}{\psi_2(|x|)} \left| \int_0^x \frac{1}{f'(\varphi(u)) \sigma(\varphi(u))} |J(u) - c_0| \, du \right| \le C$$

for all $x \in \mathbb{R}$.

Applying the assumption (i_4) we conclude that the function $[F'(x)\sigma(x)]^2$ satisfies the conditions of Theorem 4.2 for $\overline{b}(x) = 4\hat{b}(x)$. According to Theorem 4.2, the stochastic processes $\beta_T^{(2)}(t)$ converge weakly, as $T \to +\infty$, to the process

$W^* \left(\widehat{\beta}^{(1)}(t) \right)$, where the process $\widehat{\beta}^{(1)}(t)$ has the form (4.23), $W^*(t)$ is a Wiener process, and the processes $W^*(t)$ and $\widehat{\beta}^{(1)}(t)$ are independent. Consequently, the proof follows from relation (4.24).   □

Remark 4.1 Let in Eq. (2.1) $a(x) = 0$ and $\sigma(x) = 1$. If the function $g(x)$ is absolutely integrable on $\mathbb{R}$,

$$\int\limits_0^{+\infty} g(x)dx = \int\limits_0^{-\infty} g(x)dx = c_0$$

and

$$\left| \int\limits_0^x g(u)\, du - c_0 \right| \sim \frac{1}{|x|^\alpha},$$

as $|x| \to +\infty$, with $\alpha \geq \frac{1}{2}$, then the conditions of Theorem 4.5 are fulfilled.

For $\alpha > \frac{1}{2}$ the normalization in the functional $\beta_T^{(1)}(t)$ is $T^{-1/4}$ and for $\alpha = \frac{1}{2}$ it equals to $\left(\sqrt{T} \ln \sqrt{T} \right)^{-\frac{1}{2}}$.

4.2   Weak Convergence to the Functionals of the Bessel Diffusion Process

Next, we consider the behavior of the functionals $\beta_T^{(1)}(t)$ and $\beta_T^{(2)}(t)$ of the solutions ξ to Eq. (3.1).

Theorem 4.6 *Let ξ be a solution of Eq. (3.1) and the conditions of Theorem 3.3 be fulfilled. Let for some constant $C > 0$ the following inequality hold: for all x*

$$\int\limits_0^x f'(u) \left(\int\limits_0^u \frac{dv}{f'(v)} \right) du \leq C \left(1 + x^2 \right), \tag{4.25}$$

where $f'(x) = e^{-2 \int\limits_0^x a(v)\, dv}$. Let g be real-valued measurable locally integrable function, $\psi \in \Psi$ with $\alpha > 0$ and

$$q(x) := \frac{f'(x)}{\psi(|x|)} \int\limits_0^x \frac{g(u)}{f'(u)}\, du - b \, \mathrm{sign}\, x \to 0,$$

as $|x| \to +\infty$, for some constant b.

Then the stochastic processes

$$\beta_T^{(1)}(t) = \frac{1}{\sqrt{T}\psi\left(\sqrt{T}\right)} \int_0^{tT} g\left(\xi\left(s\right)\right) ds$$

converge weakly, as $T \to +\infty$, to the process

$$\beta^{(1)}(t) = 2b\left[\frac{r^{\alpha+1}(t)}{\alpha+1} - \int_0^t r^\alpha(s)\, d\widehat{W}(s)\right], \tag{4.26}$$

where $r(t) \geq 0$ is the solution of the Itô equation

$$r^2(t) = (2c+1)\, t + 2\int_0^t r(s)\, d\widehat{W}(s),$$

$c = c_0$ in case (1), $c = c_1$ in case (2), and $c = c_2$ in case (3) of Theorem 3.3.

Proof We proceed in a similar way to the proof of Theorem 4.1. Let us consider the function

$$F(x) = 2\int_0^x \left(f'(u)\int_0^u \frac{g(v)}{f'(v)}\, dv\right) du.$$

Since

$$F'(x)\, a(x) + \frac{1}{2}F''(x) = g(x)$$

a.e. with respect to the Lebesgue measure, then, applying the Itô formula (see Lemma A.3) to the process $F\left(\xi(t)\right)$, one has

$$\beta_T^{(1)}(t) = \frac{1}{\sqrt{T}\,\psi\left(\sqrt{T}\right)}\left[F\left(\xi\left(tT\right) - F\left(x_0\right)\right)\right]$$

$$-\frac{1}{\sqrt{T}\,\psi\left(\sqrt{T}\right)}\int_0^{tT} F'\left(\xi(s)\right)\, dW(s).$$

After some transformations, we have the representation

$$
\beta_T^{(1)}(t) = -\frac{F(x_0)}{\sqrt{T}\,\psi\left(\sqrt{T}\right)} + 2b \int_0^{r_T(t)} u^\alpha \, du
$$

$$
-2b \int_0^t r_T^\alpha(s)\, d\widehat{W}_T(s) + 2 \sum_{k=1}^4 S_T^{(k)}(t), \tag{4.27}
$$

where

$$
S_T^{(1)}(t) = b \int_0^{r_T(t)} \left[\frac{\psi\left(u\sqrt{T}\right)}{\psi\left(\sqrt{T}\right)} - u^\alpha \right] du,
$$

$$
S_T^{(2)}(t) = \frac{1}{\sqrt{T}\,\psi\left(\sqrt{T}\right)} \int_0^{\xi(tT)} q(u)\,\psi(|u|)\, du,
$$

$$
S_T^{(3)}(t) = -b \int_0^t \left[\frac{\psi\left(r_T(s)\sqrt{T}\right)}{\psi\left(\sqrt{T}\right)} - r_T^\alpha(s) \right] d\widehat{W}_T(s),
$$

$$
S_T^{(4)}(t) = -\int_0^t q(\xi(sT)) \frac{\psi\left(r_T(s)\sqrt{T}\right)}{\psi\left(\sqrt{T}\right)}\, dW_T(s).
$$

It follows from the proof of Theorem 3.3 that the processes $r_T(t)$ converge weakly, as $T \to +\infty$, to the process $r(t)$, which is the solution to Eq. (3.10) for $c = c_0$ in case (1), $c = c_1$ in case (2), and $c = c_2$ in case (3) of Theorem 3.3. In addition, for arbitrary constants $L > 0$ and $\varepsilon > 0$ we have

$$
\lim_{N \to +\infty} \limsup_{T \to +\infty} \mathsf{P}\left\{ \sup_{0 \le t \le L} r_T(t) > N \right\} = 0,
$$

$$
\lim_{h \to 0} \limsup_{T \to +\infty} \sup_{\substack{|t_1 - t_2| \le h \\ t_i \le L}} \mathsf{P}\left\{ |r_T(t_2) - r_T(t_1)| > \varepsilon \right\} = 0. \tag{4.28}
$$

Now we are in a position to establish that $S_T^{(k)}(t)$, $k = \overline{1,4}$, uniformly converge to zero in probability. In particular, it means that they satisfy analog of relations (4.28) as well. Let

$$
P_N = P \left\{ \sup_{0 \le t \le L} r_T(t) > N \right\}
$$

and $T_N > 0$ are introduced in Lemma A.16. For arbitrary $\varepsilon > 0$ and $L > 0$, we have the inequalities uniformly in $T \ge T_N$:

$$
P \left\{ \sup_{0 \le t \le L} \left| S_T^{(1)}(t) \right| > \varepsilon \right\} \le P_N + \frac{2}{\varepsilon} E \sup_{0 \le t \le L} \left\{ \left| S_T^{(1)}(t) \right| \chi_{r_T(t) \le N} \right\}
$$

$$
\le P_N + \frac{2}{\varepsilon} |b| \int_0^N \left| \frac{\psi \left(|u| \sqrt{T} \right)}{\psi \left(\sqrt{T} \right)} - u^\alpha \right| du,
$$

$$
P \left\{ \sup_{0 \le t \le L} \left| S_T^{(2)}(t) \right| > \varepsilon \right\} \le P \left\{ \sup_{0 \le t \le L} \left| \int_0^{\frac{\xi(tT)}{\sqrt{T}}} q \left(u\sqrt{T} \right) \frac{\psi \left(|u| \sqrt{T} \right)}{\psi \left(\sqrt{T} \right)} du \right| > \varepsilon \right\}
$$

$$
\le P_N + \frac{2}{\varepsilon} E \sup_{0 \le t \le L} \left| \int_0^{\frac{\xi(tT)}{\sqrt{T}}} q \left(u\sqrt{T} \right) \frac{\psi \left(|u| \sqrt{T} \right)}{\psi \left(\sqrt{T} \right)} du \right| \chi_{r_T(t) \le N}
$$

$$
\le P_N + \frac{2}{\varepsilon} C_N \frac{1}{\sqrt{T}} \int_{-N\sqrt{T}}^{N\sqrt{T}} |q(u)| \, du, \quad (4.29)
$$

$$
P \left\{ \sup_{0 \le t \le L} \left| S_T^{(3)}(t) \right| > \varepsilon \right\} \le P_N
$$

$$
+ 4 \left(\frac{2}{\varepsilon} \right)^2 b^2 E \int_0^L \left[\frac{\psi \left(r_T(s) \sqrt{T} \right)}{\psi \left(\sqrt{T} \right)} - r_T^\alpha(s) \right]^2 \chi_{r_T(s) \le N} ds,
$$

$$
P \left\{ \sup_{0 \le t \le L} \left| S_T^{(4)}(t) \right| > \varepsilon \right\} \le P_N
$$

$$+4\left(\frac{2}{\varepsilon}\right)^2 \mathsf{E}\int_0^L q^2\ (\xi(sT))\left[\frac{\psi\left(r_T(s)\sqrt{T}\right)}{\psi\left(\sqrt{T}\right)}\right]^2 \chi_{r_T(s)\leq N} ds$$

$$\leq P_N + 4\left(\frac{2}{\varepsilon}\right)^2 C_N^2 \mathsf{E}\int_0^L q^2\ (\xi(sT))\ \chi_{r_T(s)\leq N} ds.$$

Taking into account the convergence (see Lemma A.17) $\dfrac{\psi\left(|u|\sqrt{T}\right)}{\psi\left(\sqrt{T}\right)} - |u|^\alpha \to 0$,

as $T \to +\infty$, $|u| \neq 0$ and the boundedness of $\dfrac{\psi\left(|u|\sqrt{T}\right)}{\psi\left(\sqrt{T}\right)} - |u|^\alpha$ for $|u| \leq N$,

together with the convergence $q(x) \to 0$, as $|x| \to +\infty$, we can pass to the limit, as $T \to +\infty$, and then as $N \to +\infty$, in the inequalities for $S_T^{(k)}(t)$ with $k = 1, 2$. So, we obtain

$$\sup_{0\leq t\leq L} \left|S_T^{(k)}(t)\right| \xrightarrow{\mathsf{P}} 0, \tag{4.30}$$

as $T \to +\infty$, for $k = 1, 2$.

Now we establish a similar convergence for $S_T^{(k)}(t)$ with $k = 3, 4$. It is known that the following convergence holds (see Lemma A.17):

$$\sup_{0<\delta\leq|x|\leq N} \left|\frac{\psi\left(|x|\sqrt{T}\right)}{\psi\left(\sqrt{T}\right)} - |x|^\alpha\right| \to 0,$$

as $T \to +\infty$, for all $0 < \delta < N < +\infty$. Therefore, taking into account the monotonicity of the function $\psi(x)$, we have the convergence

$$\mathsf{E}\int_0^L \left[\frac{\psi\left(r_T(s)\sqrt{T}\right)}{\psi\left(\sqrt{T}\right)} - r_T^\alpha(s)\right]^2 \chi_{r_T(s)\leq N} ds$$

$$\leq L \sup_{0<\delta\leq|x|\leq N}\left[\frac{\psi\left(r_T(s)\sqrt{T}\right)}{\psi\left(\sqrt{T}\right)} - r_T^\alpha(s)\right]^2$$

$$+2\int_0^L\left(\left[\frac{\psi\left(\delta\sqrt{T}\right)}{\psi\left(\sqrt{T}\right)}\right]^2 + \delta^2\right) dv \to 0,$$

as $T \to +\infty$ and $\delta \to 0$.

Then we take into account the inequality (4.29) for $S_T^{(3)}(t)$ and get (4.30) for $S_T^{(3)}(t)$ as well.

Finally, in order to prove the relation (4.30) for $S_T^{(4)}(t)$, we apply the Itô formula and obtain

$$\mathsf{E} \int_0^L q^2 \left(\xi(sT) \right) \chi_{|\xi(sT)| \leq N\sqrt{T}} \, ds = \mathsf{E} \left[\Phi_T \left(\xi(LT) \right) - \Phi(x_0) \right],$$

where

$$\Phi_T(x) = \frac{1}{T} \int_0^x f'(u) \left(\int_0^u \frac{q^2(v) \chi_{|v| \leq N\sqrt{T}}}{f'(v)} \, dv \right) du.$$

It is easy to make sure that condition (4.25) implies the following convergence:

$$\lim_{|x| \to +\infty} \frac{1}{x^2} \int_0^x f'(u) \left(\int_0^u \frac{q^2(v)}{f'(v)} \, dv \right) du = 0. \tag{4.31}$$

In addition, from Theorem 3.3 we have the inequality

$$\mathsf{E} r_T(t) \leq C + C_1 t.$$

So, for an arbitrary $\varepsilon > 0$ there exists a constant $\widehat{C}_\varepsilon > 0$ such that

$$\mathsf{E} \left| \Phi_T \left(\xi(LT) \right) \right| \leq \frac{\widehat{C}_\varepsilon}{T} + \varepsilon \left(C + C_1 L \right).$$

Thus, $\mathsf{E} \left| \Phi_T \left(\xi(LT) \right) \right| \to 0$, as $T \to +\infty$. Therefore,

$$\mathsf{E} \int_0^L q^2 \left(\xi(sT) \right) \chi_{r_T(s) \leq N} \, ds \to 0,$$

as $T \to +\infty$. It means that relation (4.30) holds for $S_T^{(4)}(t)$, as well.

It is clear that the relations (4.28) hold for $S_T^{(k)}(t)$, $k = \overline{1, 4}$ and for $\widehat{W}_T(t)$ as well. It means that we can apply Skorokhod's convergent subsequence principle to the process $\left(r_T(t), \widehat{W}_T(t), S_T^{(k)}(t), k = \overline{1, 4} \right)$. Therefore, taking into account Theorem A.12 and Corollary A.3, we have that for any subsequence $T_n \to +\infty$

$$r_{T_n}(t) \xrightarrow{\mathsf{P}} r(t), \quad \widehat{W}_{T_n}(t) \xrightarrow{\mathsf{P}} \widehat{W}(t), \quad S_{T_n}^{(k)}(t) \xrightarrow{\mathsf{P}} S^{(k)}(t), \quad k = \overline{1, 4}$$

for every $t > 0$.

According to the relation (4.30), we have that $\sup\limits_{0 \le t \le L} \left| S^{(k)}_{T_n}(t) \right| \xrightarrow{P} 0$, as $T_n \to +\infty$ for $k = \overline{1,4}$ and any $L > 0$. Taking into account Theorem 3.3, we have that the processes $r(t)$ and $\widehat{W}(t)$ are related via the equation

$$r^2(t) = (2c + 1)t + 2 \int\limits_0^t r(s) \, d\widehat{W}(s), \tag{4.32}$$

where $c = c_0$ in case (1), $c = c_1$ in case (2), and $c = c_2$ in case (3) of Theorem 3.3.

Using Lemma A.7, we can go to the limit, as subsequence $T_n \to +\infty$ in the equality (4.27), and obtain that $\beta^{(1)}_{T_n}(t) \xrightarrow{P} \beta^{(1)}(t)$ for every $t \ge 0$, where

$$\beta^{(1)}(t) = 2b \left[\frac{r^{\alpha+1}}{\alpha + 1} - \int\limits_0^t r^\alpha(s) \, d\widehat{W}(s) \right], \tag{4.33}$$

and the processes $r(t)$ and $\widehat{W}(t)$ are related via Eq. (4.32).

It follows from the strong uniqueness of the solution $r(t)$ to Eq. (4.32) that the finite-dimensional distributions of the limit process $\beta^{(1)}(t)$ are unique, as well. Since the sequence $T_n \to +\infty$ is arbitrary, the finite-dimensional distributions of the processes $\beta^{(1)}_T(t)$ tend, as $T \to +\infty$, to the corresponding finite-dimensional distributions of the process $\beta^{(1)}(t)$, defined by equality (4.33). In order to establish the weak convergence of the processes $\beta^{(1)}_T(t)$ to the process $\beta^{(1)}(t)$, it is sufficient to prove the property (4.15) for these processes. It follows from the proof of Theorem 3.3 that the relation (4.15) holds for the process $r_T(t)$. According to the convergences (4.30), the property (4.15) holds for the processes $S^{(k)}_T(t)$, $k = \overline{1,4}$, as well.

In addition, using the properties of stochastic Itô integrals, we get for any $L > 0$ and $\varepsilon > 0$ the following inequalities:

$$\mathsf{P} \left\{ \sup_{|t_1 - t_2| \le h, \, t_i \le L} \left| \int\limits_0^{r_T(t_2)} u^\alpha \, du - \int\limits_0^{r_T(t_1)} u^\alpha \, du \right| > \varepsilon \right\}$$

$$\le P_N + \mathsf{P} \left\{ N^\alpha \sup_{|t_1 - t_2| \le h, \, t_i \le L} |r_T(t_2) - r_T(t_1)| > \frac{\varepsilon}{2} \right\},$$

$$\mathsf{P} \left\{ \sup_{|t_1 - t_2| \le h, \, t_i \le L} \left| \int\limits_{t_1}^{t_2} r^\alpha_T(s) \, d\widehat{W}_T(s) \right| > \varepsilon \right\} \le P_N$$

$$+\mathbf{P}\left\{4\sup_{kh\leq L}\sup_{kh\leq t\leq(k+1)h}\left|\int_{kh}^{t}r_T^{\alpha}(s)\,\chi_{r_T(s)\leq N}d\widehat{W}_T(s)\right|>\frac{\varepsilon}{2}\right\}$$

$$\leq P_N+\left(\frac{8}{\varepsilon}\right)^4\sum_{kh<L}\mathbf{E}\sup_{kh\leq t\leq(k+1)h}\left[\int_{kh}^{t}r_T^{\alpha}(s)\,\chi_{r_T(s)\leq N}d\widehat{W}_T(s)\right]^4$$

$$\leq P_N+\left(\frac{8}{\varepsilon}\right)^4\left(\frac{4}{3}\right)^4\sum_{kh<L}\mathbf{P}\sup_{kh\leq t\leq(k+1)h}\left[\int_{kh}^{(k+1)h}r_T^{\alpha}(s)\,\chi_{r_T(s)\leq N}d\widehat{W}_T(s)\right]^4$$

$$\leq P_N+\left(\frac{8}{\varepsilon}\right)^4\left(\frac{4}{3}\right)^4 36\,N^{4\alpha}\sum_{kh<L}h^2.$$

Therefore, all terms of the right-hand side in equality (4.27) satisfy the analog of the relation (4.15). So, the relation (4.15) holds for the processes $\beta_T^{(1)}(t)$, which converge weakly, as $T\to+\infty$ to the process $\beta^{(1)}(t)$, whence the proof follows. □

Remark 4.2 It follows from the proof of Theorem 4.6 that it is sufficient to have the representation (4.27) with weak convergence, as $T\to+\infty$, of the processes $r_T(t)$ to the process $r(t)$, and the convergence $\sup_{0\leq t\leq L}\left|S_T^{(k)}(t)\right|\xrightarrow{P}0$, $k=\overline{1,4}$, as $T\to+\infty$, for arbitrary $L>0$.

Moreover, Theorem 4.6 is a certain analog of Theorem 4.1 with convergence $q(x)\to 0$, as $|x|\to+\infty$. We emphasize that under the assumptions of Theorem 4.1 the function $q(x)$ may not tend to zero, as $|x|\to+\infty$, that is, this function can have "explosions" for arbitrarily large $|x|>>1$. Under additional conditions on the function $a(x)$ in Eq. (3.1) for the function $q(x)$ we can also suppose "explosions" for $|x|>>1$. In fact, we have the following theorem.

Theorem 4.7 *Let ξ be a solution to Eq. (3.1) and the conditions of Theorem 3.3 be fulfilled. Let g be a real-valued measurable locally integrable function and let $\psi\in\Psi$ with $\alpha>0$.*

Let also one of the following conditions hold

(a) *for all x, one has $f'(x)=e^{-2\int_0^x a(v)\,dv}\leq C$, where C is some constant, and*

$$\lim_{|x|\to+\infty}\frac{1}{x}\int_0^x\frac{q^2(u)}{f'(u)}\,du=0;$$

or

(b) *for all x, $0 < \delta \leq f'(x)$ and*

$$\lim_{|x| \to +\infty} \frac{f'(x)}{x} \int_0^x q^2(u)\, du = 0,$$

where

$$q(x) = \frac{f'(x)}{\psi\,(|x|)} \int_0^x \frac{g(u)}{f'(u)}\, du - b\ \text{sign}\, x,$$

and δ and b are some constants. Then the statement of Theorem 4.6 holds for the stochastic process

$$\beta_T^{(1)}(t) = \frac{1}{\sqrt{T}\,\psi\left(\sqrt{T}\right)} \int_0^{tT} g\,(\xi(s))\, ds.$$

Proof Let us obtain from (4.27) the expression of the form (4.30) for $S_T^{(k)}(t)$, $k = \overline{1,4}$, as $T \to +\infty$. The proof of this convergence for $S_T^{(1)}(t)$ and $S_T^{(3)}(t)$ is completely analogous to the proof of the same convergence in Theorem 4.6. Next, according to the conditions of Theorem 4.7, it is easy to establish convergence

$$\frac{1}{x} \int_0^x q^2(u)\, du \to 0,$$

as $|x| \to +\infty$.

The inequality (4.29) can be extended if we use the Cauchy–Schwarz inequality. Therefore,

$$P\left\{ \sup_{0 \leq t \leq L} \left| S_T^{(2)}(t) \right| > \varepsilon \right\} \leq P_N + \frac{2}{\varepsilon} C_N \frac{1}{\sqrt{T}} \int_{-N\sqrt{T}}^{N\sqrt{T}} |q(u)|\, du$$

$$\leq P_N + \frac{2}{\varepsilon} C_N\, (2N)^{1/2} \left(\frac{1}{\sqrt{T}} \int_{-N\sqrt{T}}^{N\sqrt{T}} q^2(u)\, du \right)^{\frac{1}{2}}.$$

The last inequality implies (4.30) for $S_T^{(2)}(t)$.

It is easy to obtain (4.31) from the assumptions of the theorem. Therefore, in the same way as in the proof of Theorem 4.6, we obtain (4.30) for $S_T^{(4)}(t)$ as well. Weak convergence of the processes $r_T(t)$, as $T \to +\infty$, to the processes $r(t)$ is established in Theorem 3.3. Thus, to prove Theorem 4.7 it is enough to take advantage of Remark 4.2.   □

Next, we consider certain statements for solutions ξ to Eq. (3.1), which are analogs of Theorems 4.2–4.4.

Theorem 4.8 *Let ξ be a solution of Eq. (3.1) and the conditions of Theorem 3.3 be fulfilled with $c_1 = c_2 = c_0$ and $2c_0 + 1 > 0$. Let there exist a real-valued measurable locally integrable function $g^2(x)$ such that $g(-x) = -g(x)$ and functions $\psi_i \in \Psi$ of order $\alpha_i > 0$, respectively, $i = 1, 2$, such that*

$$(1) \quad \lim_{|x| \to +\infty} \left[\frac{f'(x)}{\psi_1(|x|)} \int_0^x \frac{g^2(v)}{f'(v)} \, dv - b \, \mathrm{sign}\, x \right] = 0;$$

$$(2) \quad \left| \frac{f'(x)}{\psi_2(|x|)} \int_0^x \frac{|g(v)|}{f'(v)} \, dv \right| \leq C;$$

$$(3) \quad \lim_{T \to +\infty} \frac{\psi_2\left(\sqrt{T}\right)}{T^{\frac{1}{4}} \left(\psi_1\left(\sqrt{T}\right) \right)^{\frac{1}{2}}} = 0$$

for some constants $C > 0$ and b.
   Then the stochastic processes

$$\beta_T^{(2)}(t) = \frac{1}{T^{\frac{1}{4}} \left(\psi_1\left(\sqrt{T}\right) \right)^{\frac{1}{2}}} \int_0^{tT} g\left(\xi(s)\right) \, dW(s),$$

where the processes ξ and W are related via Eq. (3.1), converge weakly, as $T \to +\infty$, to the process $\beta^{(2)}(t) = W^*\left(\beta^{(1)}(t)\right)$, where W^* is a Wiener process, $\beta^{(1)}$ is the process from (4.33) with $\alpha = \alpha_1$, and the processes W^* and $\beta^{(1)}$ are independent.

Proof It is easy to see that

$$\beta_T^{(2)}(t) = \frac{\sqrt[4]{T}}{\sqrt{\psi_1\left(\sqrt{T}\right)}} \int_0^t g\left(r_T(s)\sqrt{T}\right) d\widehat{W}_T(s),$$

where

$$r_T(t) = \frac{|\xi(tT)|}{\sqrt{T}}, \quad \widehat{W}_T(t) = \int_0^t \operatorname{sign}\xi(sT)\, dW_T(s), \quad W_T(t) = \frac{W(tT)}{\sqrt{T}}.$$

Since $\int_0^t P\{\xi(sT) = 0\}\, ds = 0$ for all $t > 0$, $\widehat{W}_T(t)$ is an almost surely continuous martingale for every fixed $T > 0$, and its quadratic characteristic is equal to $\langle \widehat{W}_T \rangle(t) = t$. According to Doob's theorem (see [17, Chapter 1, § 1, Theorem 1]), $\widehat{W}_T(t)$ is, for every fixed $T > 0$, a Wiener process with respect to the σ-algebra $\sigma\{W_T(s),\ s \le t\}$.

Moreover, the process $\left(r_T(t),\ \widehat{W}_T(t)\right)$ satisfies Skorokhod's principle for a convergent subsequence (Theorem A.12). Hence, without loss of generality, we may assume that (see Lemma A.13), for an arbitrary subsequence $T_n \to +\infty$, the convergence $r_{T_n}(t) \xrightarrow{P} r(t)$ as well as $\widehat{W}_{T_n}(t) \xrightarrow{P} \widehat{W}(t)$ holds, as $T_n \to +\infty$, for every $t > 0$. According to Theorem 3.3, the limit processes $r(t)$ and $\widehat{W}(t)$ are related via Eq. (3.10) with $c = c_0$. Moreover, for all $L > 0$ and $\varepsilon > 0$

$$\lim_{h \to 0} \limsup_{T_n \to +\infty} P\left\{ \sup_{|t_1 - t_2| \le h;\ t_i \le L} \left| r_{T_n}(t_2) - r_{T_n}(t_1) \right| > \varepsilon \right\} = 0. \tag{4.34}$$

It is clear that a similar convergence holds for $\widehat{W}_{T_n}(t)$, as well. Let $\tau_{T_n}(t)$ be the minimal solution to the equation $\beta_{T_n}^{(1)}\left(\tau_{T_n}(t)\right) = t$, where

$$\beta_{T_n}^{(1)}(t) = \frac{\sqrt{T_n}}{\psi_1\left(\sqrt{T_n}\right)} \int_0^t g^2\left(r_{T_n}(s)\sqrt{T_n}\right) ds.$$

Note that $\tau_{T_n}(t)$ is a Markov moment with respect to the σ-algebra $\sigma\{W_{T_n}(s),\ s \le t\}$. According to Lemma A.8 we have that $\int_0^{+\infty} g^2\left(r_{T_n}(s)\sqrt{T_n}\right) ds = +\infty$ with probability 1. Thus, given an arbitrary T_n, the process $\beta_{T_n}^{(1)}(t)$ attains any positive level with probability 1. Similarly to the proof of the convergence (4.19) we obtain

$$\beta_{T_n}^{(2)}(t) - W_{T_n}^*\left(\beta^{(1)}(t)\right) \xrightarrow{P} 0, \tag{4.35}$$

as $T_n \to +\infty$, where $W_{T_n}^*(t)$ is a sequence of Wiener processes, $\beta^{(1)}(t)$ is the process from (4.33) with $\alpha = \alpha_1$.

Using properties of stochastic integrals, we get

$$\left| \mathsf{E} W_{T_n}^*(t) \widehat{W}_{T_n}(t) \right| = \left| \mathsf{E} \beta_{T_n}^{(2)} \left(\tau_{T_n}(t) \right) \widehat{W}_{T_n}(t) \right|$$

$$= \left| \mathsf{E} \frac{\sqrt[4]{T_n}}{\left(\psi_1 \left(\sqrt{T_n} \right) \right)^{\frac{1}{2}}} \int_0^{\tau_{T_n}(t)} g \left(r_{T_n}(s) \sqrt{T_n} \right) d\widehat{W}_{T_n}(s) \cdot \widehat{W}_{T_n}(t) \right|$$

$$= \frac{\sqrt[4]{T_n}}{\left(\psi_1 \left(\sqrt{T_n} \right) \right)^{\frac{1}{2}}} \left| \mathsf{E} \int_0^{t \wedge \tau_{T_n}(t)} g \left(r_{T_n}(s) \sqrt{T_n} \right) ds \right|$$

$$\leq \frac{\sqrt[4]{T_n}}{\left(\psi_1 \left(\sqrt{T_n} \right) \right)^{\frac{1}{2}}} \mathsf{E} \int_0^t \left| g \left(r_{T_n}(s) \sqrt{T_n} \right) \right| ds$$

$$= \frac{\psi_2 \left(\sqrt{T_n} \right)}{T_n^{\frac{1}{4}} \left(\psi_1 \left(\sqrt{T_n} \right) \right)^{\frac{1}{2}}} \cdot \frac{\sqrt{T_n}}{\psi_2 \left(\sqrt{T_n} \right)} \mathsf{E} \int_0^t \left| g \left(r_{T_n}(s) \sqrt{T_n} \right) \right| ds. \qquad (4.36)$$

Now we show that conditions (2) and (3) of Theorem 4.8 imply that the right-hand side of the preceding inequality tends to zero, as $T_n \to +\infty$. Consider the function

$$F(x) = 2 \int_0^x f'(u) \left(\int_0^u \frac{|g(v)|}{f'(v)} dv \right) du.$$

This function possesses a continuous derivative $F'(x)$ and an almost everywhere (with respect to the Lebesgue measure) locally integrable second derivative $F''(x)$. According to Lemma A.3, we can apply the Itô formula to the process $F(\xi(t))$ and obtain

$$\mathsf{E} \frac{\sqrt{T_n}}{\psi_2 \left(\sqrt{T_n} \right)} \int_0^t \left| g \left(r_{T_n}(s) \sqrt{T_n} \right) \right| ds = \mathsf{E} \left[\frac{F \left(\xi(tT_n) \right) - F(x_0)}{\sqrt{T_n} \, \psi_2 \left(\sqrt{T_n} \right)} \right]. \qquad (4.37)$$

According to condition (2) of Theorem 4.8, we have

$$|F(x)| \leq C \psi_2 \left(|x| \right) \cdot |x|$$

for all x for some constant $C > 0$.

Therefore,

$$
\mathsf{E}\left|\frac{F\left(\xi(tT_n)\right) - F\left(x_0\right)}{\sqrt{T_n}\,\psi_2\left(\sqrt{T_n}\right)}\right| \le C\,\mathsf{E}r_{T_n}(t)\frac{\psi_2\left(r_{T_n}(t)\sqrt{T_n}\right)}{\psi_2\left(\sqrt{T_n}\right)} \le C_1\mathsf{E}r_{T_n}^{\alpha_2+1}(t).
$$

Since the drift coefficient in Eq. (3.1) is such that $|x\,a(x)| \le C$ and $\xi(0) = x_0$, we obtain

$$
\mathsf{E}r_{T_n}^{\alpha_2+1}(t) \le C_2 + C_3 t^{\frac{\alpha_2+1}{2}}
$$

for some constants C_2 and C_3. Thus, equality (4.37) implies that

$$
\mathsf{E}\frac{\sqrt{T_n}}{\psi_2\left(\sqrt{T_n}\right)} \int\limits_0^t \left|g\left(r_{T_n}(s)\sqrt{T_n}\right)\right| ds = \widehat{C}_1 + \widehat{C}_2 t^{\frac{\alpha_2+1}{2}}.
$$

In turn, condition (3) of Theorem 4.8 and inequality (4.36) imply that

$$
\left|\mathsf{E}W_{T_n}^*(t)\widehat{W}_{T_n}(t)\right| \to 0
$$

for all $t > 0$, as $T_n \to +\infty$.

Since the processes $\widehat{W}$ and $W_{T_n}^*$ are Wiener processes and they are asymptotically uncorrelated, we conclude that $W_{T_n}^*$ does not depend asymptotically on $\widehat{W}$. The further proof of the theorem is similar to the end of the proof of Theorem 4.2.   □

Theorem 4.9 *Let ξ be a solution to Eq. (3.1) and the conditions of Theorem 3.3 be fulfilled for $c_1 = c_2 = c_0$ and $2c_0 + 1 > 0$. Let the real-valued measurable function $g^2(x)$ be locally integrable and such that*

$$
\frac{f'(x)}{|x|} \int\limits_0^x \frac{g^2(v)}{f'(v)}\, dv - b\ \mathrm{sign}\, x \to 0,
$$

as $|x| \to +\infty$, for a certain constant b.

Then the stochastic processes

$$
\beta_T^{(2)}(t) = \frac{1}{\sqrt{T}} \int\limits_0^{tT} g\left(\xi(s)\right) dW(s)
$$

converge weakly, as $T \to +\infty$, to the process $\sqrt{b\,(2c_0 + 1)}W^*(t)$, where W^* is a Wiener process.

Proof It is clear that

$$\beta_T^{(2)}(t) = \int\limits_0^t g\left(\xi(sT)\right) dW_T(s)$$

with probability 1 for every $t > 0$. Let $\tau_T(t)$ be the minimal solution to the equation $\beta_T^{(1)}(\tau_T(t)) = t$, where

$$\beta_T^{(1)}(t) = \int\limits_0^t g^2\left(\xi(sT)\right) ds.$$

Since $\tau_T(t)$ is a Markov moment with respect to the σ-algebra $\sigma\{W(sT), s \le t\}$ for every fixed $T > 0$ and, according to Lemma A.8, the equality

$$\int\limits_0^{+\infty} g^2\left(r_{T_n}(s)\sqrt{T_n}\right) ds = +\infty$$

holds with probability 1, then for every fixed $T > 0$ the process $\beta_T^{(2)}(\tau_T(t))$ is a Wiener process. Denote this process by $W_T^*(t)$. Thus,

$$\beta_T^{(2)}(\tau_T(t)) = W_T^*\left(\beta_T^{(1)}(t)\right).$$

It is easy to see that the function $g^2(x)$ satisfies the conditions of Theorem 4.6 for $\alpha = 1$. Therefore, according to Theorem 4.6, the process $\beta_T^{(1)}(t)$ converges weakly, as $T \to +\infty$, to the process

$$\beta^{(1)}(t) = 2b\left[\frac{r^2(t)}{2} - \int\limits_0^t r(s) d\widehat{W}(s)\right] = b\left(2c_0 + 1\right) t.$$

It follows from the condition of Theorem 4.9 that $b \ge 0$. Since the limit process for $\beta_T^{(1)}(t)$ is degenerate, then convergence of $\beta_T^{(1)}(t)$ to $\beta^{(1)}(t)$, as $T \to +\infty$, for every $t > 0$ in probability, follows from the weak convergence of $\beta_T^{(1)}(t)$ to $\beta^{(1)}(t)$, as $T \to +\infty$.

Consequently, similarly to the proof of the relation (4.19), we have

$$\beta_T^{(2)}(t) - W_T^*\left(\beta^{(1)}(t)\right) \xrightarrow{P} 0 \qquad (4.38)$$

as $T \to +\infty$. Since the process

$$\frac{W_T^* \left(\beta^{(1)}(t) \right)}{\sqrt{b(2c_0 + 1)}}$$

is a Wiener process for $b > 0$, according to (4.38), we have the convergence of the finite-dimensional distributions of the process $\beta_T^{(2)}$ to the corresponding distributions of the process $\sqrt{b\,(2c_0 + 1)}\,W^*$, where W^* is a Wiener process.

The proof of the relation (4.15) for the process $\beta_T^{(2)}$ coincides completely with the proof of this convergence for the process $\beta_T^{(2)}$ in Theorem 4.2. To complete the proof, it remains to use Theorem A.13.   □

Theorem 4.10 *Let ξ be a solution of Eq. (3.1) and the conditions of Theorem 3.3 be fulfilled for $c_1 = c_2 = c_0$ and $2c_0 + 1 > 0$. Assume that a real-valued measurable function $g(x)$ is locally integrable and there are two constants a and b for which*

$$(1) \qquad \frac{1}{|x|} \int_0^x f'(u) \left(\int_a^u \frac{g(v)}{f'(v)}\, dv \right) du \to 0,$$

and

$$(2) \qquad \frac{f'(x)}{|x|} \int_0^x f'(u) \left[\int_a^u \frac{g(v)}{f'(v)}\, dv \right]^2 du - b\, \operatorname{sign} x \to 0,$$

as $|x| \to +\infty$. Then the stochastic processes

$$\beta_T^{(1)}(t) = \frac{1}{\sqrt{T}} \int_0^{tT} g\left(\xi(s) \right)\, ds$$

converge weakly, as $T \to +\infty$, to the process

$$2\sqrt{b\,(2c_0 + 1)}\,W^*(t),$$

where W^* is a Wiener process.

Proof Consider the function

$$F(x) = 2 \int_0^x f'(u) \left(\int_a^u \frac{g(v)}{f'(v)}\, dv \right) du.$$

This continuous function possesses a continuous derivative $F'(x)$ and an almost everywhere (with respect to the Lebesgue measure) locally integrable second derivative $F''(x)$. According to Lemma A.3, we can apply the Itô formula to the process $F(\xi(t))$. Using the equality

$$F'(x)a(x) + \frac{1}{2}F''(x) = g(x)$$

a.e. with respect to the Lebesgue measure, we obtain that

$$\beta_T^{(1)}(t) = \frac{F(\xi(tT)) - F(x_0)}{\sqrt{T}} - \frac{1}{\sqrt{T}}\int_0^{tT} F'(\xi(s))\,dW(s) \tag{4.39}$$

with probability 1 for all $t \geq 0$ and for every T. It follows from condition (1) of Theorem 4.10 that $\frac{F(x)}{x} \to 0$, as $|x| \to +\infty$. In addition, taking into account the inequality (3.16), we obtain

$$\mathsf{E}\sup_{0 \leq t \leq L} \frac{\xi^2(tT)}{T} \leq C_L.$$

It is easy to establish that (see the proof of Theorem 3.3)

$$\sup_{0 \leq t \leq L}\left|\frac{F(\xi(tT))}{\sqrt{T}}\right| \xrightarrow{\mathsf{P}} 0,$$

as $T \to +\infty$. So, for every $t \geq 0$

$$\sup_{0 \leq t \leq L}\left|\beta_T^{(1)}(t) + \frac{1}{\sqrt{T}}\int_0^{tT} F'(\xi(s))\,dW(s)\right| \xrightarrow{\mathsf{P}} 0,$$

as $T \to +\infty$. Since

$$\frac{1}{\sqrt{T}}\int_0^{tT} F'(\xi(s))\,dW(s) = \int_0^t F'(\xi(sT))\,dW_T(s) = W_T^*\left(\int_0^t \left[F'(\xi(sT))\right]^2 ds\right),$$

where W_T^* is a Wiener process for every T (see the proof of Theorem 4.7), then

$$\sup_{0 \leq t \leq L}\left|\beta_T(t) + W_T^*\left(\int_0^t \left[F'(\xi(sT))\right]^2 ds\right)\right| \xrightarrow{\mathsf{P}} 0,$$

as $T \to +\infty$. According to condition (2), the function $\left[F'(x)\right]^2$ satisfies assumptions of Theorem 4.6 with $\alpha = 1$. That is why the processes

$$\beta_T^{(1)}(t) = \int\limits_0^t \left[F'\left(\xi(sT)\right)\right]^2 ds$$

converge weakly, as $T \to +\infty$, to the degenerate process

$$\beta^{(1)}(t) = 4b\left(2c_0 + 1\right)t.$$

The rest of the proof is similar to the corresponding proof of Theorem 4.9.   □

4.3   Results About Weak Convergence of the Mixed Functionals

Next, we consider the asymptotic behavior, as $t \to +\infty$, of the distributions of functionals of the following form:

$$I(t) = F\left(\xi(t)\right) + \int\limits_0^t g\left(\xi(s)\right) dW(s), \tag{4.40}$$

where $F = F(x)$ is a continuous functions, $g = g(x)$ is a real-valued measurable function, $x \in \mathbb{R}$, g is locally square integrable, the processes ξ and W are related via Eq. (2.1) with $\sigma(x) \equiv 1$ and $\xi(0) = x_0$. Let us now consider the behavior of these functionals depending on the solutions of two classes of equations.

Definition 4.2 Let K_1 be the class of equations of the form (2.1) with $\sigma(x) \equiv 1$, $\xi(0) = x_0$, and such a that

$$\sup_{x \in \mathbb{R}} \left| \int\limits_0^x a(v) \, dv \right| \le C$$

for a certain constant $C > 0$, and convergence (2.13) holds for

$$f(x) = \int\limits_0^x \exp\left\{-2\int\limits_0^u a(v)\, dv\right\} du.$$

Also, let K_2 be the class of equations of the form (2.1) with $\sigma(x) \equiv 1, \xi(0) = x_0$, and such a that

$$\sup_{x \in \mathbb{R}} |x\, a(x)| \le C, \quad \text{and} \quad \lim_{|x| \to +\infty} \frac{1}{x} \int_0^x v\, a(v)\, dv = c_0 > -\frac{1}{2}.$$

According to Lemma 2.1 and Theorem 3.1, the solutions ξ of equations belonging to classes $K_i,\, i = 1, 2$ are stochastically unstable. It is clear that the functional $\beta_T^{(2)}(t)$ is a functional of the form (4.40). It follows from (4.5) that the functional $\beta_T^{(1)}(t)$ is a functional of the form (4.40), as well. In addition, using the Itô formula, we have equality

$$\int_0^t g\, (\xi(s))\, d\xi(s) = \int_0^t g\, (\xi(s))\, a\, (\xi(s))\, ds + \int_0^t g\, (\xi(s))\, dW(s)$$

$$= \Phi\, (\xi(t)) - \Phi\, (x_0) + \int_0^t \left[g\, (\xi(s)) - \Phi'\, (\xi(s)) \right] dW(s), \quad (4.41)$$

where

$$\Phi\, (x) = 2 \int_0^x f'(u) \left(\int_0^u \frac{g(v)\, a(v)}{f'(v)} dv \right) du.$$

Thus, the functional $\int_0^t g\, (\xi(s))\, d\xi(s)$ is a functional of the form (4.40). Therefore, using the results obtained earlier about the functionals $\beta_T^{(1)}(t)$ and $\beta_T^{(2)}(t)$, it is easy to get similar results for properly normalized functionals from (4.40).

Here we present only the formulation of the following theorems.

Theorem 4.11 *Let ξ be a solution of Eq. (2.1) belonging to the class K_1. Let two functions F and g define the functional $I(t)$ by equality (4.40) and let the function $\psi \in \Psi$ be a regularly varying function of order $\alpha \ge 0$. Assume that*

(1) $\quad \lim_{|x| \to +\infty} \left[\dfrac{F(x)}{f(x)\, \psi\, (|f(x)|)} - \bar{a}(x) \right] = 0, \quad \bar{a}(x) = \begin{cases} a_1, & x \ge 0, \\ a_2, & x < 0. \end{cases}$

(2) $\quad \left| \dfrac{g(x)}{f'(x)\, \psi\, (|f(x)|)} \right| \chi_{(|x| \le N)} \le C_N \ \text{ for any } \ N > 0 \ \text{ and}$

$\qquad \lim_{|x| \to +\infty} \left[\dfrac{g(x)}{f'(x)\, \psi\, (|f(x)|)} - \bar{b}(x) \right] = 0, \quad \bar{b}(x) = \begin{cases} b_1, & x \ge 0, \\ b_2, & x < 0 \end{cases}$

for some constants $C_N > 0$ and $a_i, b_i,\, i = 1, 2$.

Then the stochastic processes

$$I_T(t) = \frac{I(tT)}{\sqrt{T}\,\psi\left(\sqrt{T}\right)}$$

converge weakly, as $T \to +\infty$, to the process

$$I_0(t) = \overline{a}\,(\zeta(t))\,\zeta(t)\,|\zeta(t)|^{\alpha} + \int_0^t \overline{b}\,(\zeta(s))\,|\zeta(s)|^{\alpha}\,d\zeta(s),$$

where ζ is the solution of Eq. (2.8).

Remark 4.3

(1) If $\alpha = 0, \overline{a}(x) = a_0 \text{ sign } x$, and $\overline{b}(x) = -a_0 \text{ sign } x$, then

$$I_0(t) = a_0\left[|\zeta(t)| - \int_0^t \text{sign }\zeta(s)\,d\zeta(s)\right] = a_0 L_\zeta^0(t),$$

where $L_\zeta^0(t)$ is the local time at zero of the process ζ on the interval $[0, t]$.

(2) If $\alpha = 1, \overline{a}(x) = a_0 \text{ sign } x$, and $\overline{b}(x) = -2a_0 \text{ sign } x$, then the Itô formula for the process $\zeta^2(t)$ yields

$$I_0(t) = a_0 \int_0^t \overline{\sigma}^2\,(\zeta(s))\,ds,$$

and for $\sigma_1 = \sigma_2 = \sigma_0$ we have $I_0(t) = a_0\sigma_0^2 t$.

Theorem 4.12 *Let ξ be a solution of Eq. (2.1) belonging to the class K_1. Let two functions F and g define the functional $I(t)$ by equality (4.40) and let the functions $\psi_i \in \Psi$ be regularly varying functions of orders $\alpha_i \geq 0, i = 1, 2$. Assume that*

(1) $$\lim_{|x|\to+\infty}\left[\frac{F(x)}{\sqrt{|f(x)|}\,\psi_1\,(|f(x)|)} - \overline{a}(x)\right] = 0, \quad \overline{a}(x) = \begin{cases} a_1, & x > 0, \\ a_2, & x < 0. \end{cases}$$

(2) $$\lim_{|x|\to+\infty}\left[\frac{1}{\psi_1\,(|f(x)|)}\int_0^x \frac{g^2\,(|v|)}{f'\,(v)}\,dv - \overline{b}(x)\right] = 0,$$

$$\overline{b}(x) = \begin{cases} b_1, & x \geq 0, \\ b_2, & x < 0, \end{cases} \quad b_1 \geq 0, \quad b_2 \leq 0.$$

$$(3) \quad \left| \frac{1}{\psi_2 \left(|f(x)| \right)} \int\limits_0^x \frac{|g(v)|}{f'(v)} \, dv \right| \le C, \quad \lim_{T \to +\infty} \frac{\psi_2 \left(\sqrt{T} \right)}{T^{\frac{1}{4}} \left(\psi_1 \left(\sqrt{T} \right) \right)^{\frac{1}{2}}} = 0$$

for some constants $C > 0$ and a_i, b_i, $i = 1, 2$.
   Then the stochastic processes

$$I_T(t) = \frac{I(tT)}{T^{\frac{1}{4}} \left(\psi_1 \left(\sqrt{T} \right) \right)^{\frac{1}{2}}}$$

converge weakly, as $T \to +\infty$, to the process

$$I_0^*(t) = \overline{a} \left(\zeta(t) \right) |\zeta(t)|^{\frac{1+\alpha_1}{2}} + W^* \left(\beta^{(1)}(t) \right),$$

where

$$\beta^{(1)}(t) = 2 \left[\int\limits_0^{\zeta(t)} \overline{b}(x) \, |x|^{\alpha_1} \, dx - \int\limits_0^t \overline{b} \left(\zeta(s) \right) |\zeta(s)|^{\alpha_1} \, d\zeta(s) \right],$$

ζ is a solution of Eq. (2.8), W^* is a Wiener process, and the processes W^* and ζ are independent.

Remark 4.4

(1) If $\alpha_1 = 0, \overline{b}(x) = b_0 \operatorname{sign} x$, then

$$I_0^*(t) = \overline{a} \left(\zeta(t) \right) \sqrt{|\zeta(t)|} + W^* \left(2b_0 L_\zeta^0(t) \right),$$

   where $L_\zeta^0(t)$ is the local time at zero of the process ζ on the interval $[0, t]$, W^* is a Wiener process, and the processes W^* and ζ are independent.
(2) If $\alpha_1 = 1, \overline{b}(x) = b_0 \operatorname{sign} x$, then

$$I_0^*(t) = \overline{a} \left(\zeta(t) \right) |\zeta(t)| + W^* \left(b_0 \int\limits_0^t \overline{\sigma}^2 \left(\zeta(s) \right) \, ds \right),$$

   where the processes W^* and ζ are independent.
   In particular, if $\sigma_1 = \sigma_2 = \sigma_0$, then

$$I_0^*(t) = \overline{a} \left(\zeta(t) \right) |\zeta(t)| + W^* \left(b_0 \sigma_0^2 t \right).$$

(3) If $\overline{b}(x) \equiv 0$, then $I_0^*(t) = \overline{a} \left(\zeta(t) \right) |\zeta(t)|^{\frac{1+\alpha_1}{2}}$, $\alpha_1 \ge 0$.

Theorem 4.13 *Let ξ be a solution of Eq. (3.1) belonging to the class K_2. Let two functions $F = F(x)$ and $g = g(x)$ define the functional $I(t)$ by equality (4.40) and let $\psi = \psi(r) \in \Psi$ be a regularly varying function of order $\alpha > 0$. Assume that*

$$(1) \quad \lim_{|x| \to +\infty} \frac{F(x)}{|x|\ \psi\ (|x|)} = a_0;$$

$$(2) \quad \text{the function} \ \frac{g(x)}{\psi\ (|x|)} \ \text{is bounded and} \quad \lim_{|x| \to +\infty} \left[\frac{g(x)}{\psi\ (|x|)} - b_0 \ sign\ x \right] = 0,$$

where a_0 and b_0 are some constants.
   Then the stochastic processes

$$I_T(t) = \frac{I(tT)}{\sqrt{T} \psi_1 \left(\sqrt{T} \right)}$$

converge weakly, as $T \to +\infty$, to the process

$$I_0(t) = a_0 r^{\alpha+1}(t) + b_0 \int_0^t r^\alpha(s)\, d\widehat{W}(s),$$

where the stochastic process r and the Wiener process $\widehat{W}$ are related via Eq. (4.32) with $c = c_0$.

Remark 4.5 If $\alpha = 1$, then

$$I_0(t) = \left(a_0 + \frac{b_0}{2} \right) r^2(t) - \frac{b_0}{2} (2c_0 + 1)\, t.$$

In particular, in the case $a_0 = -\frac{b_0}{2}$ we have that $I_0(t) = a_0\, (2c_0 + 1)\, t.$

Theorem 4.14 *Let ξ be a solution of Eq. (3.1) belonging to the class K_2. Let two functions $F = F(x)$ and $g = g(x)$ define the functional $I(t)$ by equality (4.40). Assume that there are two regularly varying functions $\psi_i = \psi_i(r) \in \Psi$ of orders $\alpha_i > 0$, $i = 1, 2$, and an odd locally square integrable function $\widehat{g} = \widehat{g}(x)$ such that*

$$(1) \quad \lim_{|x| \to +\infty} \frac{F(x)}{\sqrt{|x|}\ \psi_1\ (|x|)} = a_0;$$

$$(2) \quad \lim_{|x| \to +\infty} \left[\frac{f'(x)}{\psi_1\ (|x|)} \int_0^x \frac{\widehat{g}^2(u)}{f'(u)}\, du - b_0 \ sign\ x \right] = 0;$$

$$(3) \quad \lim_{|x| \to +\infty} \left[\frac{f'(x)}{\psi_1(|x|)} \int_0^x \frac{[g(u) - \widehat{g}(u)]^2}{f'(u)} \, du \right] = 0;$$

$$(4) \quad \left| \frac{f'(x)}{\psi_2(|x|)} \int_0^x \frac{|g(u)|}{f'(u)} \, du \right| \le C, \quad \lim_{T \to +\infty} \frac{\psi_2\left(\sqrt{T}\right)}{T^{\frac{1}{4}} \left(\psi_1\left(\sqrt{T}\right)\right)^{\frac{1}{2}}} = 0$$

for some constants $C > 0$ and a_0, b_0.
   Then the stochastic processes

$$I_T(t) = \frac{I(tT)}{T^{\frac{1}{4}} \left(\psi_1\left(\sqrt{T}\right)\right)^{\frac{1}{2}}}$$

converge weakly, as $T \to +\infty$, to the process

$$I_0^*(t) = a_0 \left[r(t)\right]^{\frac{1+\alpha_1}{2}} + W^*\left(\beta^{(1)}(t)\right),$$

where

$$\beta^{(1)}(t) = 2b_0 \left[\frac{r^{\alpha_1 + 1}(t)}{\alpha_1 + 1} - \int_0^t r^{\alpha_1}(s) \, d\widehat{W}(s) \right].$$

Here $r \ge 0$ is a solution of Eq. (4.32) with $c = c_0$, W^* is a Wiener process, and the processes W^* and $\widehat{W}$ are independent.

Remark 4.6 Condition (4) of Theorem 4.14 is used only when we prove that W^* and $\widehat{W}$ are independent. If $\alpha_1 = 1$ and $a_0 = 0$, then we can omit condition (4) and get $I_0^*(t) = \sqrt{b_0 (2c_0 + 1)} \, W^*(t)$.

4.4   Examples

Example 4.1 Let the coefficients $a(x)$ and $\sigma(x)$ of Eq. (2.1) satisfy the sufficient conditions of Theorem 2.1. Let the real-valued measurable function $g(x)$ be locally integrable.

(1) If the function $g(x) \left[f'(x)\sigma^2(x)\right]^{-1}$ is absolutely integrable on $\mathbb{R}$ and

$$\lambda := \int_{\mathbb{R}} \frac{g(x)}{f'(x)\sigma^2(x)} \, dx = 0,$$

then the stochastic processes

$$\beta_T^{(1)}(t) = \frac{1}{\sqrt{T}} \int_0^{tT} g\left(\xi(s)\right) ds$$

converge weakly, as $T \to +\infty$, to the process $\beta^{(1)}(t) \equiv 0$.

Indeed, in this case the conditions of Theorem 4.1 are fulfilled with $\psi(r) \equiv 1$, $b_1 = b_2 = b_0$, where

$$b_0 = \int_0^{+\infty} \frac{g(x)}{f'(x)\sigma^2(x)} dx.$$

Therefore, according to Theorem 4.1 the stochastic processes $\beta_T^{(1)}(t)$ converge weakly, as $T \to +\infty$, to the process $\beta^{(1)}(t)$ of the form (4.4) with $\alpha = 0$ and $\overline{b}(x) = b_0$. So, in this case, the limit process $\beta^{(1)}(t) \equiv 0$.

Note, that for $\lambda = 0$ the behavior, as $t \to +\infty$, of the functional $\int_0^t g\left(\xi(s)\right) ds$ depends on the rate of convergence $\alpha(x) \to 0$, as $|x| \to +\infty$, where

$$\alpha(x) = \int_0^x \frac{g(u)}{f'(u)\sigma^2(u)} du - b_0.$$

For example, if

$$\lim_{|x| \to +\infty} \frac{1}{\sqrt{|f(x)|}} \int_0^x f'(u)\alpha(u) du = 0 \quad \text{and} \quad \int_{\mathbb{R}} f'(x)\alpha^2(u) du < +\infty,$$

then the stochastic processes

$$\beta_T^{(1)}(t) = \frac{1}{\sqrt[4]{T}} \int_0^{tT} g\left(\xi(s)\right) ds$$

converge weakly, as $T \to +\infty$, to the process $W^*\left(\beta^{(1)}(t)\right)$, where W^* is a Wiener process, $\beta^{(1)}$ is a process of the form (4.4) with $\alpha = 0$,

$$b_1 = 4 \int_0^{+\infty} f'(u)\alpha^2(u) du, \quad b_2 = 4 \int_0^{-\infty} f'(u)\alpha^2(u) du,$$

and the processes W^* and $\beta^{(1)}$ are independent.

Indeed, it follows from the representation

$$\int\limits_0^t g\,(\xi(s))\,ds = F\,(\xi(t)) - F\,(\xi(0)) - \int\limits_0^t F'\,(\xi(s))\,\sigma\,(\xi(s))\,dW(s),$$

where $F(x) = 2\int\limits_0^x f'(u)\,\alpha(u)\,du$ and from the proof of Theorem 4.5. Note, that the function $F'(x)\sigma(x)$ satisfies the conditions of Theorem 4.2.

(2) If $\lambda \neq 0$, then the conditions of Theorem 4.1 are fulfilled with $\psi(r) \equiv 1$, $\alpha = 0$,

$$b_1 = \int\limits_0^{+\infty} g(u)\left[f'(u)\,\sigma^2(u)\right]^{-1}du, \text{ and } b_2 = \int\limits_0^{-\infty} g(u)\left[f'(u)\,\sigma^2(u)\right]^{-1}du.$$

Therefore, according to Theorem 4.1, the stochastic processes

$$\beta_T^{(1)}(t) = \frac{1}{\sqrt{T}}\int\limits_0^{tT} g\,(\xi(s))\,ds$$

converge weakly, as $T \to +\infty$, to the process

$$\beta^{(1)}(t) = 2\left[\int\limits_0^{\zeta(t)}\overline{b}(u)\,du - \int\limits_0^t \overline{b}(\zeta(s))\,d\zeta(s)\right],$$

where $\overline{b}(x) = b_1$ for $x > 0$ and $\overline{b}(x) = b_2$ for $x < 0$, $\zeta(t)$ is the solution to Eq. (2.8). In particular, if $b_2 = -b_1$, then

$$\beta^{(1)}(t) = 2b_1\left[|\zeta(t)| - \int\limits_0^t \text{sign}\,\zeta(s)\,d\zeta(s)\right] = 2b_1 L_\zeta^0(t),$$

where $L_\zeta^0(t)$ is the local time at zero of the process ζ on the interval $[0,\,t]$.

Example 4.2 Let in Eq. (2.1) the coefficients $a(x) = 0$ and $\sigma(x) = 1$ for all x.

(1) If $g(x) = \sin x$, then the conditions of Theorem 4.4 are fulfilled with $c_0 = 1$ and $\sigma_0^2 = \frac{1}{2}$. Therefore, the stochastic processes

$$\frac{1}{\sqrt{T}}\int\limits_0^{tT}\sin W(s)\,ds$$

converge weakly, as $T \to +\infty$, to the process $\sqrt{2}W^*(t)$, where W^* is a Wiener process.

In addition, in this case the conditions of Theorem 4.3 are fulfilled with $\sigma_0^2 = \frac{1}{2}$. Therefore, the stochastic processes

$$\frac{1}{\sqrt{T}} \int_0^{tT} \sin W(s)\, dW(s)$$

converge weakly, as $T \to +\infty$, to the process $\frac{1}{\sqrt{2}} W^*(t)$ with a Wiener process W^*.

(2) If

$$g(x) = \frac{\overline{c}(x) \sin x}{\sqrt{1 + x^2}}, \quad \overline{c}(x) = \begin{cases} c_1, & x \geq 0, \\ c_2, & x < 0, \end{cases}$$

then the conditions of Theorem 4.2 are fulfilled with

$$\psi_1(r) \equiv 1, \quad \psi_2(r) \equiv \ln(1+r), \quad b_1 = c_1^2 \int_0^{+\infty} \frac{\sin^2 x}{1 + x^2}\, dx, \quad b_2 = c_2^2 \int_0^{-\infty} \frac{\sin^2 x}{1 + x^2}\, dx.$$

Therefore, according to Theorem 4.2, the stochastic processes

$$\beta_T^{(2)}(t) = \frac{1}{\sqrt[4]{T}} \int_0^{tT} \frac{\overline{c}(W(s)) \sin W(s)}{\sqrt{1 + W^2(s)}}\, dW(s)$$

converge weakly, as $T \to +\infty$, to the process $W^* \left(\beta^{(1)}(t) \right)$, where $W^*(t)$ is a Wiener process,

$$\beta^{(1)}(t) = 2 \left[\int_0^{W(t)} \overline{b}(u)\, du - \int_0^t \overline{b}(W(s))\, dW(s) \right],$$

$\overline{b}(x) = b_1$ for $x > 0$ and $\overline{b}(x) = b_2$ for $x < 0$. Furthermore, the Wiener processes $W^*(t)$ and $W(t)$ are independent. In particular, if $c_1^2 = c_2^2$, then

$$\beta^{(1)}(t) = 2b_1 L_W^0(t).$$

(3) If the function $g(x)$ is such that

$$\int_0^x g(u)\, du - \operatorname{sign} x = \begin{cases} \sin(x - n)n^2\pi, & |x| \in \left(n, \, n + \frac{1}{n^2}\right), \\ 0, & |x| \notin \left(n, \, n + \frac{1}{n^2}\right), \end{cases}$$

then for this function the conditions of Theorem 4.1 are fulfilled with $\psi(r) \equiv 1$ and $\bar{b}(x) = \text{sign}\,x$, and $\alpha = 0$. Note, that $q(x) \nrightarrow 0$, as $|x| \to +\infty$ and

$$\frac{1}{x}\int_0^x q^2(u)\,du \to 0,$$

as $|x| \to +\infty$. Then the stochastic processes $\frac{1}{\sqrt{T}}\int_0^{tT} g(W(s))\,ds$ converge weakly, as $T \to +\infty$, to the process

$$\beta^{(1)}(t) = 2\left[|W(t)| - \int_0^t \text{sign}\,W(s)\,dW(s)\right] = 2L_W^0(t).$$

Example 4.3 Let the coefficients of Eq. (2.1) be $a(x) \equiv 0$ and $\sigma(x) \equiv 1$. Let the initial condition of Eq. (2.1) be $\xi(0) = 0$. The solution of Eq. (2.8) has the form $\zeta(t) = W(t)$.

In this case $f(x) = x$, $\varphi(x) = x$, and

$$J(x) = \int_0^x \frac{g(\varphi(u))}{[f'(\varphi(u))\,\sigma(\varphi(u))]^2}\,du = \int_0^x g(u)\,du.$$

Let the equality $J(x) = \frac{\sin x}{\sqrt{1+x^2}}$ be fulfilled, that is,

$$g(x) = \left(\frac{\sin x}{\sqrt{1+x^2}}\right)' = \frac{\cos x}{\sqrt{1+x^2}} - \frac{x\sin x}{\left(1+x^2\right)}$$

and

$$F(x) = 2\int_0^{f(x)} J(u)\,du = 2\int_0^x \frac{\sin u}{\sqrt{1+u^2}}\,du.$$

The conditions of the Theorem 4.5 are satisfied with $\psi_1(|x|) = 1$, $(\alpha_1 = 0)$, $c_0 = 0$, $\hat{b}(x) = b_0\,\text{sign}\,x$,

$$b_0 = \int_0^{+\infty} \frac{\sin u}{\sqrt{1+u^2}}\,du,$$

$\psi_2(|x|) = \ln\left(1+|x|^2\right)$, $(\alpha_2 = 0)$.

According to Theorem 4.5 the stochastic processes

$$\beta_T^{(1)}(t) = \frac{1}{\sqrt[4]{T}} \int_0^{tT} g\left(\xi(s)\right) ds$$

converge weakly, as $T \to +\infty$, to the process $W^*\left(\widehat{\beta}^{(1)}(t)\right)$, where W^* is a Wiener process,

$$\widehat{\beta}^{(1)}(t) = 8b_0^2 \left[\int_0^{W(t)} \operatorname{sign} u \, du - \int_0^t \operatorname{sign}\left(W(s)\right) dW(s) \right] = 8b_0^2 L_W^0(t),$$

the Wiener process W^* and the process $\widehat{\beta}^{(1)}$ are independent.

Example 4.4 Let in Eq. (3.1) $a(x) = \frac{c_0 x}{1+x^2}$ for all x and $2c_0 + 1 > 0$.

(1) If $g(x) = \sin x$ and $c_0 = 1$, then the conditions of Theorem 4.10 are fulfilled for $a = 0$ and $b = \frac{1}{6}$. Therefore, the stochastic processes

$$\frac{1}{\sqrt{T}} \int_0^{tT} \sin \xi(s) \, ds$$

converge weakly, as $T \to +\infty$, to the process $\sqrt{2}W^*(t)$, where W^* is a Wiener process.

In addition, the conditions of Theorem 4.9 are fulfilled for $b = \frac{1}{6}$. Therefore, according to Theorem 4.9 the stochastic processes

$$\frac{1}{\sqrt{T}} \int_0^{tT} \sin \xi(s) \, dW(s),$$

where the processes $\xi(t)$ and $W(t)$ are related via Eq. (2.1), converge weakly, as $T \to +\infty$, to the process $\frac{1}{\sqrt{2}} W^*(t)$, where W^* is a Wiener process.

(2) If $g(x) = \frac{\sin x}{\sqrt[8]{1+x^2}}$ and $c_0 = -\frac{1}{4}$, then the conditions of Theorem 4.8 are fulfilled for $\psi_1(r) = \sqrt{r} \ln r$, $\psi_2(r) = r^{\frac{3}{4}}$, $b = \frac{1}{2}$. Therefore, the stochastic processes

$$\frac{1}{T^{3/8}\sqrt{\ln \sqrt{T}}} \int_0^{tT} \frac{\sin \xi(s)}{\sqrt[8]{1+\xi^2(s)}} \, dW(s),$$

where the processes $\xi(t)$ and $W(t)$ are related via Eq. (2.1), converge weakly, as $T \to +\infty$, to the process $W^* \left(\beta^{(1)}(t)\right)$, where W^* is a Wiener process,

$$\beta^{(1)}(t) = \frac{2}{3}r^{3/2}(t) - \int_0^t \sqrt{r(s)}\, d\widehat{W}(s), \quad r^2(t) = \frac{1}{2}t + \int_0^t r(s)\, d\widehat{W}(s),$$

and the Wiener processes $\widehat{W}$ and W^* are independent.

Example 4.5 Let in Eq. (2.1) $a(x) = \frac{x}{1+x^4}$ and $\sigma(x) \equiv 1$. In this case, according to Theorem 2.1, the solution to Eq. (2.8) is $\zeta(t) = \sigma_0 W(t)$, where $\sigma_0 = e^{-\frac{\pi}{2}}$.

(1) Let $g(x) \to g_0$, as $|x| \to +\infty$. Using the equality (4.41), we obtain that the conditions of Theorem 4.11 are fulfilled for the functional $\int_0^t g\left(\xi(s)\right)\, d\xi(s)$ with $\psi\left(|x|\right) \equiv 1, \overline{a}(x) = 2\overline{\alpha}(x), \overline{b}(x) = g_0\sigma_0^{-1} - 2\overline{\alpha}(x)$, where

$$\overline{\alpha}(x) = \begin{cases} \alpha_1, & x \geq 0, \\ \alpha_2, & x < 0, \end{cases} \quad \alpha_1 = \int_0^{+\infty} \frac{g(v)a(v)}{f'(v)}\, dv, \quad \alpha_2 = \int_0^{-\infty} \frac{g(v)a(v)}{f'(v)}\, dv.$$

Therefore, the stochastic processes

$$\frac{1}{\sqrt{T}} \int_0^{tT} g\left(\xi(s)\right)\, d\xi(s)$$

converge weakly, as $T \to +\infty$, to the process

$$I_0(t) = 2\sigma_0 \left[\overline{\alpha}\left(W(t)\right) W(t) - \int_0^t \overline{\alpha}\left(W(s)\right)\, dW(s) \right] + g_0 W(t).$$

In particular,

(a) if $\alpha_0 = \alpha_1 = -\alpha_2$, then $I_0(t) = 2\alpha_0\sigma_0 L_W^0(t) + g_0 W(t)$;
(b) if $\alpha_0 = \alpha_1 = \alpha_2$, then $I_0(t) = g_0 W(t)$.

(2) Let

$$I(t) = |\xi(t)| - \int_0^t \operatorname{sign}\xi(s)\, d\xi(s) = L_\xi^0(t)$$

be the local time at zero of the process ξ on the interval $[0, t]$.

Using the equality (4.41), we obtain that the conditions of Theorem 4.11 are fulfilled for the functional $I(t)$ with $\psi(|x|) \equiv 1$, $\overline{a}(x) = \text{sign}\,x$, $\overline{b}(x) = -\text{sign}\,x$. Therefore, the stochastic processes $\frac{1}{\sqrt{T}}L_\xi^0(tT)$ converge weakly, as $T \to +\infty$, to the process $I_0(t) = \sigma_0 L_W^0(t)$.

Example 4.6 Let in Eq. (3.1) $a(x) = \frac{x}{1+x^2}$. The conditions of Theorem 4.14 are fulfilled for the functional $\int_0^t \sin\xi(s)\,d\xi(s)$ with $\psi_1(|x|) = |x|$, $a_0 = 0$, $\alpha_1 = 1$, $\widehat{g}(x) = \sin x$, $b_0 = \frac{1}{6}$, $c_0 = 1$. So, according to Remark 4.6, the stochastic processes

$$\frac{1}{\sqrt{T}} \int\limits_0^{tT} \sin(\xi(s))\,d\xi(s)$$

converge weakly, as $T \to +\infty$, to the process $I_0^*(t) = \frac{1}{\sqrt{2}}W^*(t)$, where W^* is a Wiener process.

Chapter 5
Asymptotic Behavior of Homogeneous Additive Functionals Defined on the Solutions of Itô SDEs with Non-regular Dependence on a Parameter

In this chapter, we consider homogeneous one-dimensional stochastic differential equations with non-regular dependence on a parameter. The asymptotic behavior of the mixed functionals of the form $I_T(t) = F_T(\xi_T(t)) + \int_0^t g_T(\xi_T(s)) \, d\xi_T(s), t \geq 0$ is studied as $T \to +\infty$. Here ξ_T is a strong solution to the stochastic differential equation $d\xi_T(t) = a_T(\xi_T(t)) \, dt + dW_T(t)$, $T > 0$ is a parameter, $a_T = a_T(x)$ is a set of measurable functions, $|a_T(x)| \leq L_T$ for all $x \in \mathbb{R}$, $W_T = W_T(t)$ are standard Wiener processes, $F_T = F_T(x)$, $x \in \mathbb{R}$ are continuous functions, and $g_T = g_T(x)$, $x \in \mathbb{R}$ are locally bounded real-valued functions. Section 5.1 contains some preliminary remarks. We prove a theorem about the weak compactness of the family of some processes in Sect. 5.2. Section 5.3 includes a theorem concerning the weak convergence of some stochastic processes to the solutions of Itô SDEs. In Sect. 5.4 we consider the asymptotic behavior of integral functionals of Lebesgue integral type. Section 5.5 is devoted to asymptotic behavior of the integral functionals of martingale type. The explicit form of the limiting processes for $I_T(t)$ is established in Sect. 5.6 under very non-regular dependence of g_T and a_T on the parameter T. This section summarizes the main results and their proofs. Section 5.7 contains several examples. Auxiliary results are collected in Sect. 5.8.

5.1 Preliminaries

Consider the following Itô stochastic differential equation:

$$d\xi_T(t) = a_T(\xi_T(t)) \, dt + dW_T(t), \ t \geq 0, \ \xi_T(0) = x_0, \qquad (5.1)$$

© Springer Nature Switzerland AG 2020
G. Kulinich et al., *Asymptotic Analysis of Unstable Solutions of Stochastic Differential Equations*, Bocconi & Springer Series 9,
https://doi.org/10.1007/978-3-030-41291-3_5

where $T > 0$ is a parameter, $a_T = a_T(x)$, $x \in \mathbb{R}$ are real-valued measurable functions, such that for some constants $L_T > 0$ and for all $x \in \mathbb{R}$ $|a_T(x)| \leq L_T$, and $W_T = \{W_T(t), t \geq 0\}$, $T > 0$ is a family of standard Wiener processes defined on a complete probability space $(\Omega, \mathfrak{F}, P)$.

It is known from Theorem 4 in [82] that for any $T > 0$ and $x_0 \in \mathbb{R}$ Eq. (5.1) possesses a unique strong pathwise solution $\xi_T = \{\xi_T(t), t \geq 0\}$ and this solution is a homogeneous strong Markov process.

We suppose that the drift coefficient $a_T(x)$ in Eq. (5.1) can have a very non-regular dependence on the parameter. For example, the drift coefficient can be a "δ"-type sequence at some points x_k, as $T \to +\infty$. Otherwise, it can be equal to $\sqrt{T} \sin\left((x - x_k)\sqrt{T}\right)$ or it can have degeneracies of some other types.

It is known from [17, § 16] that the asymptotic behavior of the solution ξ_T of Eq. (5.1) is closely related to the asymptotic behavior of harmonic functions, i.e., functions satisfying the following ordinary differential equation a.e. with respect to the Lebesgue measure:

$$f_T'(x)a_T(x) + \frac{1}{2} f_T''(x) = 0.$$

It is obvious that the functions $f_T(x)$ have the form

$$f_T(x) = c_T^{(1)} \int_0^x \exp\left\{-2 \int_0^u a_T(v)\, dv\right\} du + c_T^{(2)}, \tag{5.2}$$

where $c_T^{(1)}$ and $c_T^{(2)}$ are some families of constants.

The latter functions possess the continuous derivatives $f_T'(x)$ and their second derivatives $f_T''(x)$ exist a.e. with respect to the Lebesgue measure and are locally integrable. Note that $c_T^{(1)}$ play the role of normalizing constants and $c_T^{(2)}$ of centering constants, respectively, in the limit theorems (see [81, §6]). Further, for simplicity, we assume that $c_T^{(1)} \equiv 1$, $c_T^{(2)} \equiv 0$.

In this chapter our assumption concerning the coefficient $a_T(x)$ of Eq. (5.1) is that there exists a family of functions $G_T = G_T(x)$, $x \in \mathbb{R}$ with continuous derivatives $G_T'(x)$ and locally integrable second derivatives $G_T''(x)$ a.e. with respect to the Lebesgue measure, such that for all $T > 0$ and $x \in \mathbb{R}$ the following inequalities hold true:

(A_1) $\left(G_T'(x)a_T(x) + \frac{1}{2} G_T''(x)\right)^2 + \left(G_T'(x)\right)^2 \leq C\left(1 + (G_T(x))^2\right),$
$|G_T(x_0)| \leq C.$

Suppose additionally that the unique strong solution of Eq. (5.1) satisfies the following assumptions:

(A_2) $\lim\limits_{N \to +\infty} \limsup\limits_{T \to +\infty} P\left\{\sup\limits_{0 \leq t \leq L} |\xi_T(t)| > N\right\} = 0$ for any constant $L > 0$.

(A_3) There exist constants $\delta > 0$, $C > 0$, and $m_1 \geq 0$ such that the following inequality holds:

$$\left| \int_0^x f_T'(u) \left| \int_0^u \frac{1}{|f_T'(v)|^{1+\delta}} \, dv \right|^{\frac{1}{1+\delta}} du \right| \leq C \left[1 + |G_T(x)|^{m_1} \right]$$

for all $x \in \mathbb{R}$ and $T > 0$, where $f_T'(x) = \exp\left\{ -2 \int_0^x a_T(v) \, dv \right\}$.

(A_4) There exist a bounded function $\psi(|x|)$, $x \in \mathbb{R}$ and a constant $m_2 \geq 0$ such that $\psi(|x|) \to 0$ as $|x| \to 0$, and for all $x \in \mathbb{R}$, $T > 0$ and any measurable bounded set B the following inequality holds:

$$\int_0^x f_T'(u) \left(\int_0^u \frac{\chi_B(G_T(v))}{f_T'(v)} \, dv \right) du \leq \psi(\lambda(B)) \left[1 + |G_T(x)|^{m_2} \right],$$

where $\chi_B(v)$ is the indicator function of a set B and $\lambda(B)$ is the Lebesgue measure of B.

Let $K(G_T)$ be the class of equations of the form (5.1) whose coefficients $a_T(x)$ and solution satisfy conditions (A_1)–(A_4). It is easy to understand that the class $K(G_T)$ does not depend on the constants $c_T^{(1)}$ and $c_T^{(2)}$ in the representation (5.2).

It is clear that if there exist constants $\delta > 0$ and $C > 0$ such that $0 < \delta \leq f_T'(x) \leq C$ for all $x \in \mathbb{R}$, $T > 0$, then such equations (5.1) belong to the class $K(G_T)$ for $G_T(x) = f_T(x)$. We denote this subclass as K_1. Note that the class $K(G_T)$ contains in particular the equations for which at some points x_k we have a convergence $f_T'(x_k) \to +\infty$ or a convergence $f_T'(x_k) \to 0$, as $T \to +\infty$. For example, consider Eq. (5.1) with $a_T(x) = \frac{c_0 T x}{1+x^2 T}$. It is easy to obtain that $f_T'(x) = \frac{1}{(1+x^2 T)^{c_0}}$ and if $c_0 > -\frac{1}{2}$, such equations belong to the class $K(G_T)$ with $G_T(x) = x^2$ (here at point $x \neq 0$ we have $f_T'(x) \to 0$ for $c_0 > 0$, $f_T'(x) \to +\infty$ for $-\frac{1}{2} < c_0 < 0$, as $T \to +\infty$, and $f_T'(x) \equiv 1$ for $c_0 = 0$).

For the class of equations $K(G_T)$ we study the asymptotic behavior, as $T \to +\infty$, of the distributions of the following functionals:

$$\beta_T^{(1)}(t) = \int_0^t g_T(\xi_T(s)) \, ds, \quad \beta_T^{(2)}(t) = \int_0^t g_T(\xi_T(s)) \, dW_T(s),$$

$$I_T(t) = F_T(\xi_T(t)) + \int_0^t g_T(\xi_T(s)) \, dW_T(s), \quad \beta_T(t) = \int_0^t g_T(\xi_T(s)) \, d\xi_T(s),$$

where the processes ξ_T and W_T are related via Eq. (5.1), g_T is a family of measurable, locally bounded real-valued functions, F_T is a family of continuous real-valued functions.

Assume that for certain locally bounded functions $q_T(x)$ and any constant $N > 0$ the following conditions hold:

(A_5) $\qquad \psi_T^{(1)}(x) := f_T'(x) \int\limits_0^x \frac{q_T(v)}{f_T'(v)}\, dv \to 0$, a.e., as $T \to +\infty$,

herewith $|\psi_T^{(1)}(x)| \chi_{|x| \le N} \le C_N$ for arbitrary $N > 0$;

(A_6) $\qquad$ for measurable and locally bounded functions $g_T(x)$ and $g_0(x)$

$$\psi_T^{(2)}(x) := f_T'(x) \int\limits_0^x \frac{g_T(v)}{f_T'(v)}\, dv - g_0\,(G_T(x))\, G_T'(x) \to 0, \text{ a.e., as } T \to +\infty,$$

herewith $|\psi_T^{(2)}(x)| \chi_{|x| \le N} \le C_N$;

(A_7) $\qquad$ there exist some constants $C > 0$ and $\alpha_i \ge 0$, $i = 1, 2$ such that

$$\left| f_T'(x) \int\limits_0^x \frac{q_T(v)}{f_T'(v)}\, dv \right| \le C\,[1 + |x|^{\alpha_1}], \qquad |G_T(x)| \ge C|x|^{\alpha_2}$$

for all $x \in \mathbb{R}$.

5.2 Theorem Concerning the Weak Compactness

In what follows we denote by C, L, N, C_N any constants which do not depend on T and x.

Theorem 5.1 *Let ξ_T be a solution to Eq. (5.1) and there exists a family of functions $G_T(x)$, which satisfies assumption (A_1). Then the family of the processes $\zeta_T = \{\zeta_T(t) = G_T(\xi_T(t)), t \ge 0\}$ is weakly compact.*

Proof The functions $G_T(x)$ have continuous derivatives $G_T'(x)$ for all $T > 0$, the second derivatives $G_T''(x)$ exist a.e. with respect to the Lebesgue measure and are locally integrable. Therefore (Lemma A.3) we can apply the Itô formula to the process $\zeta_T(t) = G_T(\xi_T(t))$, and with probability 1, for all $t \ge 0$, we obtain

$$\zeta_T(t) = G_T(x_0) + \int\limits_0^t L_T(\xi_T(s))\, ds + \int\limits_0^t G_T'(\xi_T(s))\, dW_T(s), \qquad (5.3)$$

where

$$L_T(x) = G_T'(x)\, a_T(x) + \frac{1}{2} G_T''(x).$$

Let $\chi_N(t) = \chi\left\{\underset{0 \le s \le t}{\sup} |\zeta_T(s)| \le N\right\}$. It is clear that for $s \le t$ we have $\chi_N(t)\chi_N(s) = \chi_N(t)$ with probability 1. Thus, according to (5.3), the following equality holds with probability 1:

$$\zeta_T(t)\chi_N(t) = \zeta_T(0)\chi_N(t)$$
$$+\chi_N(t)\int_0^t L_T(\xi_T(s))\chi_N(s)\,ds + \chi_N(t)\int_0^t G'_T(\xi_T(s))\chi_N(s)\,dW_T(s). \tag{5.4}$$

Hence, using condition (A_1) and the properties of stochastic integrals, we obtain that

$$\mathsf{E}\zeta_T^2(t)\chi_N(t) \le 3\left[\mathsf{E}\zeta_T^2(0)\chi_N(t) + \mathsf{E}\left(\int_0^t L_T(\xi_T(s))\chi_N(s)\,ds\right)^2\right.$$

$$\left. + \mathsf{E}\left(\int_0^t G'_T(\xi_T(s))\chi_N(s)\,dW_T(s)\right)^2\right]$$

$$\le 3\left[\mathsf{E}\zeta_T^2(0)\chi_N(t) + t\int_0^t \mathsf{E}L_T^2(\xi_T(s))\chi_N(s)\,ds + \int_0^t \mathsf{E}\left[G'_T(\xi_T(s))\right]^2 \chi_N(s)\,ds\right]$$

$$\le 3\left[C + t\int_0^t C\left[1 + \mathsf{E}\zeta_T^2(s)\chi_N(s)\right]\,ds + C\int_0^t \left[1 + \mathsf{E}\zeta_T^2(s)\chi_N(s)\right]\,ds\right]$$

$$\le C^{(1)} + C^{(2)}\int_0^t \mathsf{E}\zeta_T^2(s)\chi_N(s)\,ds, \tag{5.5}$$

where $C^{(1)} = 3C(1+t+t^2)$, $C^{(2)} = 3C(1+t)$, $C > 0$ is a constant from condition (A_1), $0 \le t \le L$.

Using the Gronwall inequality, we conclude that there exists a constant K_L, not depending on T, and such that for $0 \le t \le L$

$$\mathsf{E}\zeta_T^2(t)\chi_N(t) \le K_L.$$

Let $N \uparrow +\infty$, then $\zeta_T^2(t)\chi_N(t) \uparrow \zeta_T^2(t)$, and we get the inequality

$$\underset{0 \le t \le L}{\sup} \mathsf{E}\zeta_T^2(t) \le K_L. \tag{5.6}$$

Similarly to (5.5), using (5.3) and the inequality

$$\mathsf{E} \sup_{0 \le t \le L} \left[\int_0^t G'_T(\xi_T(s)) \, dW_T(s) \right]^2 \le 4 \int_0^L \mathsf{E} \left[G'_T(\xi_T(s)) \right]^2 \, ds,$$

we conclude that

$$\mathsf{E} \sup_{0 \le t \le L} |\zeta_T(t)|^2 \le 3 \left[G_T^2(x_0) + L \int_0^L C \left[1 + \mathsf{E} \zeta_T^2(s) \right] ds + \int_0^L C \left[1 + \mathsf{E} \zeta_T^2(s) \right] ds \right]$$

$$\le \widetilde{C}_L^{(1)} + \widetilde{C}_L^{(2)} \int_0^L \mathsf{E} \left[\zeta_T(s) \right]^2 \, ds.$$

Therefore, considering (5.6), we obtain the inequality

$$\mathsf{E} \sup_{0 \le t \le L} |\zeta_T(t)|^2 \le \widetilde{K}_L \tag{5.7}$$

for all $L > 0$, where the constants $\widetilde{K}_L$ do not depend on T.

Using the inequalities for martingales and for stochastic integrals (see. [17, Part I, § 3, Theorem 6]), we obtain that

$$\mathsf{E} \sup_{0 \le t \le L} \left| \int_0^t G'_T(\xi_T(s)) \chi_N(s) \, dW_T(s) \right|^{2m}$$

$$\le \left(\frac{2m}{2m-1} \right)^{2m} \mathsf{E} \left| \int_0^L G'_T(\xi_T(s)) \chi_N(s) \, dW_T(s) \right|^{2m}$$

$$\le \left(\frac{2m}{2m-1} \right)^{2m} [m(2m-1)]^{m-1} L^{m-1} \int_0^L \mathsf{E} \left[G'_T(\xi_T(s)) \right]^{2m} \chi_N(s) \, ds,$$

for any natural number m. Therefore, similarly to (5.7), we have the inequality

$$\mathsf{E} \sup_{0 \le t \le L} |\zeta_T(t)|^{2m} \le K_{L,m}. \tag{5.8}$$

Furthermore, for all $\alpha > 0$ there exists $m \in \mathbf{N}$ such that $\alpha \leq 2m$ and, for a random variable η, we have

$$\mathsf{E}|\eta|^{\alpha} \leq 1 + \mathsf{E}|\eta|^{2m}.$$

The last inequality together with (5.8) implies that

$$\mathsf{E} \sup_{0 \leq t \leq L} |\zeta_T(t)|^{\alpha} \leq K_{L,\alpha} \tag{5.9}$$

for all $\alpha > 0$ and $L > 0$, where the constants $K_{L,\alpha}$ do not depend on T.

We have that for $t_1 < t_2 \leq L$

$$\mathsf{E}\left[\zeta_T(t_2) - \zeta_T(t_1)\right]^4$$

$$\leq 8 \left[\mathsf{E}\left(\int_{t_1}^{t_2} L_T(\xi_T(s))\, ds \right)^4 + \mathsf{E}\left(\int_{t_1}^{t_2} G'_T(\xi_T(s))\, dW_T(s) \right)^4 \right]$$

$$\leq 8 \left[(t_2 - t_1)^3 \int_{t_1}^{t_2} \mathsf{E}\,|L_T(\xi_T(s))|^4\, ds + 36(t_2 - t_1) \int_{t_1}^{t_2} \mathsf{E}\left[G'_T(\xi_T(s)) \right]^4\, ds \right].$$

Therefore, considering condition (A_1) and inequality (5.9), we get

$$\mathsf{E}\left[\zeta_T(t_2) - \zeta_T(t_1)\right]^4 \leq C_L |t_2 - t_1|^2, \tag{5.10}$$

where the constants C_L do not depend on T.

According to (5.6) and (5.10), we have the relations of weak compactness:

$$\begin{aligned}
&\lim_{N \to +\infty} \limsup_{T \to +\infty} \sup_{0 \leq t \leq L} \mathsf{P}\{|\zeta_T(t)| > N\} = 0, \\
&\lim_{h \to 0} \limsup_{T \to +\infty} \sup_{|t_1 - t_2| \leq h,\ t_i \leq L} \mathsf{P}\{|\zeta_T(t_2) - \zeta_T(t_1)| > \varepsilon\} = 0
\end{aligned} \tag{5.11}$$

for any $L > 0$, $\varepsilon > 0$.

It means that we can apply Skorokhod's convergent subsequence principle (see Theorem A.12) for the process $\zeta_T(t)$ for all $0 \leq t \leq L$. According to this principle, given an arbitrary sequence $T'_n \to +\infty$, we can choose a subsequence $T_n \to +\infty$, a probability space $(\tilde{\Omega}, \tilde{\mathfrak{F}}, \tilde{\mathsf{P}})$, and a stochastic processes $\tilde{\zeta}_{T_n}(t)$ and $\zeta(t)$, defined on this space, such that their finite-dimensional distributions coincide with those of the processes $\zeta_{T_n}(t)$, and, moreover, $\tilde{\zeta}_{T_n}(t) \xrightarrow{\tilde{\mathsf{P}}} \zeta(t)$, as $T_n \to +\infty$, for all $0 \leq t \leq L$. The processes $\tilde{\zeta}_{T_n}(t)$ and $\zeta(t)$ can be assumed to be separable.

Using (5.10), we have

$$\mathsf{E}\left[\tilde{\zeta}_{T_n}(t_2) \doteq \tilde{\zeta}_{T_n}(t_1)\right]^4 \le C_L |t_2 - t_1|^2$$

for all $0 \le t_1 \le t_2 \le L$.

By Fatou's lemma,

$$\mathsf{E}\left[\zeta(t_2) - \zeta(t_1)\right]^4 \le C_L |t_2 - t_1|^2.$$

Thus, the processes $\tilde{\zeta}_{T_n}(t)$ and $\zeta(t)$ are continuous with probability 1. We have that the finite-dimensional distributions of the processes $\tilde{\zeta}_{T_n}(t)$ converge, as $T_n \to +\infty$, to the correspondent finite-dimensional distributions of the process $\zeta(t)$. For the weak convergence of the processes $\zeta_{T_n}(t)$ it is sufficient (see [19, Chapter IX, § 2]) to prove that

$$\lim_{h \to 0} \limsup_{T_n \to +\infty} \mathsf{P}\left\{ \sup_{|t_1 - t_2| \le h, \, t_i \le L} |\zeta_{T_n}(t_2) - \zeta_{T_n}(t_1)| > \varepsilon \right\} = 0 \tag{5.12}$$

for any $L > 0$, $\varepsilon > 0$.

In order to do this, we use the Hölder and Burkholder–Gundy inequalities and get

$$\mathsf{P}\left\{ \sup_{|t_1 - t_2| \le h, \, t_i \le L} |\zeta_{T_n}(t_2) - \zeta_{T_n}(t_1)| > \varepsilon \right\}$$

$$\le \sum_{kh \le L} \mathsf{P}\left\{ \sup_{kh \le t \le (k+1)h} |\zeta_{T_n}(t) - \zeta_{T_n}(kh)| > \frac{\varepsilon}{4} \right\}$$

$$\le \left(\frac{4}{\varepsilon}\right)^4 8 \sum_{kh \le L} \left\{ \mathsf{E} \sup_{kh \le t \le (k+1)h} \left(\int_{kh}^t L_{T_n}(\xi_{T_n}(s))\, ds \right)^4 \right.$$

$$+ \mathsf{E} \sup_{kh \le t \le (k+1)h} \left. \left(\int_{kh}^t G'_{T_n}(\xi_{T_n}(s))\, dW_{T_n}(s) \right)^4 \right\} \le \left(\frac{4}{\varepsilon}\right)^4 K_L \sum_{kh \le L} (h^4 + h^2),$$

$$\tag{5.13}$$

where K_L do not depend on T_n. Obviously, (5.13) implies (5.12). The proof of Theorem 5.1 is complete.   □

5.3   Weak Convergence to the Solutions of Itô SDEs

In this section we obtain sufficient conditions for the weak convergence of some stochastic processes to the solutions of Itô SDEs.

Theorem 5.2 *Let ξ_T be a solution of Eq. (5.1) belonging to the class $K(G_T)$ and $G_T(x_0) \to y_0$, as $T \to +\infty$. Assume that there exist measurable locally bounded functions $a_0(x)$ and $\sigma_0(x)$ such that:*

(1) the functions

$$q_T^{(1)}(x) = G_T'(x) \, a_T(x) + \frac{1}{2} G_T''(x) - a_0(G_T(x)),$$

$$q_T^{(2)}(x) = \left[G_T'(x) \right]^2 - \sigma_0^2(G_T(x)),$$

satisfy assumption (A_5);

(2) the Itô equation

$$\zeta(t) = y_0 + \int_0^t a_0(\zeta(s)) \, ds + \int_0^t \sigma_0(\zeta(s)) \, d\widehat{W}(s) \tag{5.14}$$

has a unique weak solution $\left(\zeta, \widehat{W} \right)$.

Then the stochastic processes $\zeta_T = G_T(\xi_T(\cdot))$ converge weakly, as $T \to +\infty$, to the solution ζ of Eq. (5.14).

Proof Rewrite Eq. (5.3) as

$$\zeta_T(t) = G_T(x_0) + \int_0^t a_0(\zeta_T(s)) \, ds + \alpha_T^{(1)}(t) + \eta_T(t), \tag{5.15}$$

where

$$\alpha_T^{(1)}(t) = \int_0^t q_T^{(1)}(\xi_T(s)) \, ds, \quad q_T^{(1)}(x) = G_T'(x) a_T(x) + \frac{1}{2} G_T''(x) - a_0(G_T(x)),$$

$$\eta_T(t) = \int_0^t G_T'(\xi_T(s)) \, dW_T(s).$$

The functions $q_T^{(1)}(x)$ satisfy the conditions of Lemma 5.2. Thus, for any $L > 0$

$$\sup_{0 \le t \le L} \left| \alpha_T^{(1)}(t) \right| \xrightarrow{P} 0, \tag{5.16}$$

as $T \to +\infty$. It is clear that $\eta_T(t)$ is a family of continuous martingales with quadratic characteristics

$$\langle \eta_T \rangle (t) = \int_0^t \left(G_T'(\xi_T(s)) \right)^2 ds = \int_0^t \sigma_0^2(\xi_T(s)) \, ds + \alpha_T^{(2)}(t), \tag{5.17}$$

where

$$\alpha_T^{(2)}(t) = \int_0^t q_T^{(2)}(\xi_T(s)) \, ds, \quad q_T^{(2)}(x) = \left(G_T'(x) \right)^2 - \sigma_0^2 \left(G_T(x) \right).$$

The functions $q_T^{(2)}(x)$ satisfy the conditions of Lemma 5.2. Thus, for any $L > 0$

$$\sup_{0 \le t \le L} \left| \alpha_T^{(2)}(t) \right| \xrightarrow{P} 0, \tag{5.18}$$

as $T \to +\infty$.

We have that relations (5.10) and (5.11) hold for the processes $\zeta_T(t)$ and $\eta_T(t)$. According to convergence (5.16) and (5.18), these relations hold for the processes $\alpha_T^{(k)}(t)$, $k = 1, 2$ as well. It means that we can apply Skorokhod's convergent subsequence principle (see Theorem A.12) for the process

$$\left(\zeta_T(t), \eta_T(t), \alpha_T^{(1)}(t), \alpha_T^{(2)}(t) \right).$$

According to this principle, given an arbitrary sequence $T_n' \to +\infty$, we can choose a subsequence $T_n \to +\infty$, a probability space $(\widetilde{\Omega}, \widetilde{\mathfrak{F}}, \widetilde{P})$, and a stochastic process

$$\left(\widetilde{\zeta}_{T_n}(t), \widetilde{\eta}_{T_n}(t), \widetilde{\alpha}_{T_n}^{(1)}(t), \widetilde{\alpha}_{T_n}^{(2)}(t) \right)$$

defined on this space and such that its finite-dimensional distributions coincide with those of the process

$$\left(\zeta_{T_n}(t), \eta_{T_n}(t), \alpha_{T_n}^{(1)}(t), \alpha_{T_n}^{(2)}(t) \right)$$

and, moreover,

$$\widetilde{\zeta}_{T_n}(t) \overset{\widetilde{\mathsf{P}}}{\to} \widetilde{\zeta}(t), \quad \widetilde{\eta}_{T_n}(t) \overset{\widetilde{\mathsf{P}}}{\to} \widetilde{\eta}(t), \quad \widetilde{\alpha}_{T_n}^{(1)}(t) \overset{\widetilde{\mathsf{P}}}{\to} \widetilde{\alpha}^{(1)}(t), \quad \widetilde{\alpha}_{T_n}^{(2)}(t) \overset{\widetilde{\mathsf{P}}}{\to} \widetilde{\alpha}^{(2)}(t)$$

for all $0 \le t \le L$, where $\widetilde{\zeta}(t), \widetilde{\eta}(t), \widetilde{\alpha}^{(1)}(t), \widetilde{\alpha}^{(2)}(t)$ are some stochastic processes.

Evidently, relations (5.16) and (5.18) imply that $\widetilde{\alpha}^{(k)}(t) \equiv 0, k = 1, 2$ a.s. According to (5.10), the processes $\widetilde{\zeta}(t)$ and $\widetilde{\eta}(t)$ are continuous. Moreover, applying Lemma 5.5 together with equalities (5.15) and (5.17), we obtain that

$$\widetilde{\zeta}_{T_n}(t) = G_{T_n}(x_0) + \int_0^t a_0(\widetilde{\zeta}_{T_n}(s))\, ds + \widetilde{\alpha}_{T_n}^{(1)}(t) + \widetilde{\eta}_{T_n}(t)$$

and   (5.19)

$$\langle \widetilde{\eta}_{T_n} \rangle(t) = \int_0^t \sigma_0^2(\widetilde{\zeta}_{T_n}(s))\, ds + \widetilde{\alpha}_{T_n}^{(2)}(t),$$

where $\widetilde{\zeta}_{T_n}(t) \overset{\widetilde{\mathsf{P}}}{\to} \widetilde{\zeta}(t), \widetilde{\eta}_{T_n}(t) \overset{\widetilde{\mathsf{P}}}{\to} \widetilde{\eta}(t)$ and $\sup_{0 \le t \le L} \left| \widetilde{\alpha}_{T_n}^{(k)}(t) \right| \overset{\widetilde{\mathsf{P}}}{\to} 0, k = 1, 2$ as $T_n \to +\infty$.

In addition, an analog of the convergence (5.12) holds for the processes $\widetilde{\zeta}_{T_n}(t)$ and $\widetilde{\eta}_{T_n}(t)$. Therefore,

$$\sup_{0 \le t \le L} \left| \widetilde{\zeta}_{T_n}(t) - \widetilde{\zeta}(t) \right| \overset{\widetilde{\mathsf{P}}}{\to} 0, \qquad \sup_{0 \le t \le L} \left| \widetilde{\eta}_{T_n}(t) - \widetilde{\eta}(t) \right| \overset{\widetilde{\mathsf{P}}}{\to} 0,$$

as $T_n \to +\infty$. According to Lemma 5.3 we can pass to the limit in (5.19) and obtain

$$\widetilde{\zeta}(t) = y_0 + \int_0^t a_0(\widetilde{\zeta}(s))\, ds + \widetilde{\eta}(t),$$ (5.20)

where $\widetilde{\eta}(t)$ is a continuous martingale with the quadratic characteristics

$$\langle \widetilde{\eta} \rangle(t) = \int_0^t \sigma_0^2(\widetilde{\zeta}(s))\, ds.$$

Now, it is well known that the latter representation provides the existence of a Wiener process $\widehat{W}(t)$ such that

$$\widetilde{\eta}(t) = \int_0^t \sigma_0(\widetilde{\zeta}(s))\, d\widehat{W}(s).$$ (5.21)

Thus, the process $\left(\tilde{\zeta}(t),\ \widehat{W}(t)\right)$ satisfies Eq. (5.14) and the processes $\tilde{\zeta}_{T_n}(t)$ converge weakly, as $T_n \to +\infty$, to the process $\tilde{\zeta}(t)$. Since the subsequence $T_n \to +\infty$ is arbitrary and since the solution of Eq. (5.14) is weakly unique, the proof of Theorem 5.2 is complete. □

5.4   Asymptotic Behavior of Integral Functionals of the Lebesgue Integral Type

In this section we obtain the sufficient conditions for the weak convergence of some integral functionals of the Lebesgue integral type.

Theorem 5.3 *Let ξ_T be a solution of Eq. (5.1) belonging to the class $K(G_T)$ and let the assumptions of Theorem 5.2 hold. Assume that for measurable and locally bounded functions g_T there exists a measurable and locally bounded function g_0 such that the functions*

$$q_T(x) = g_T(x) - g_0(G_T(x))$$

satisfy assumption (A_5). Then the stochastic processes

$$\beta_T^{(1)}(t) = \int_0^t g_T(\xi_T(s))\, ds$$

converge weakly, as $T \to +\infty$, to the process

$$\beta^{(1)}(t) = \int_0^t g_0(\zeta(s))\, ds,$$

where ζ is the solution of Eq. (5.14).

Proof It is clear that, for all $t > 0$, with probability 1

$$\beta_T^{(1)}(t) = \int_0^t g_0(\zeta_T(s))\, ds + \alpha_T(t),$$

where

$$\alpha_T(t) = \int_0^t q_T(\xi_T(s))\, ds, \qquad q_T(x) = g_T(x) - g_0(G_T(x)).$$

The functions $q_T(x)$ satisfy the conditions of Lemma 5.2. Thus, for any $L > 0$

$$\sup_{0 \le t \le L} |\alpha_T(t)| \overset{\text{P}}{\to} 0,$$

as $T \to +\infty$. Similarly to (5.19), we obtain the equality

$$\widetilde{\beta}_{T_n}^{(1)}(t) = \int_0^t g_0(\widetilde{\zeta}_{T_n}(s)) \, ds + \widetilde{\alpha}_{T_n}(t), \tag{5.22}$$

where

$$\widetilde{\zeta}_{T_n}(t) \overset{\widetilde{\text{P}}}{\to} \widetilde{\zeta}(t) \quad \text{and} \quad \sup_{0 \le t \le L} |\widetilde{\alpha}_{T_n}(t)| \overset{\widetilde{\text{P}}}{\to} 0,$$

as $T_n \to +\infty$. The process $\widetilde{\zeta}(t)$ is a solution to Eq. (5.20), whereas by Lemma 5.5 the finite-dimensional distributions of the stochastic process $\beta_{T_n}^{(1)}(t)$ coincide with those of the process $\widetilde{\beta}_{T_n}^{(1)}(t)$.

Using Lemma 5.3 and equality (5.22) we conclude that

$$\sup_{0 \le t \le L} \left| \widetilde{\beta}_{T_n}^{(1)}(t) - \int_0^t g_0(\widetilde{\zeta}(s)) \, ds \right| \overset{\widetilde{\text{P}}}{\to} 0$$

as $T_n \to +\infty$. Thus, the process $\beta_{T_n}^{(1)}(t)$ converges weakly as $T_n \to +\infty$ to the process $\beta^{(1)}(t) = \int_0^t g_0(\zeta(s)) \, ds$, where ζ is a solution of Eq. (5.14). Since the subsequence $T_n \to +\infty$ is arbitrary and since a solution ζ of Eq. (5.14) is weakly unique, the proof of Theorem 5.3 is complete. $\square$

Theorem 5.4 *Let ξ_T be a solution of Eq. (5.1) belonging to the class $K(G_T)$, and let the assumptions of Theorem 5.2 hold. Assume that, for measurable and locally bounded functions g_T, there exists a measurable locally bounded function g_0 such that assumption (A_6) holds. Then the stochastic processes*

$$\beta_T^{(1)}(t) = \int_0^t g_T(\xi_T(s)) \, ds$$

converge weakly, as $T \to +\infty$, to the process

$$\widetilde{\beta}^{(1)}(t) = 2 \left(\int_{y_0}^{\zeta(t)} g_0(x) \, dx - \int_0^t g_0(\zeta(s)) \, \sigma_0(\zeta(s)) \, d\widehat{W}(s) \right),$$

where ζ and the Wiener process $\widehat{W}$ are related via Eq. (5.14).

Proof Consider the function

$$\Phi_T(x) = 2 \int\limits_0^x f'_T(u) \left(\int\limits_0^u \frac{g_T(v)}{f'_T(v)} \, dv \right) du.$$

Applying the Itô formula to the process $\Phi_T(\xi_T(t))$, where $\xi_T(t)$ is a solution to Eq. (5.1), we get that

$$
\begin{aligned}
\beta_T^{(1)}(t) &= \Phi_T(\xi_T(t)) - \Phi_T(x_0) - \int_0^t \Phi'_T(\xi_T(s)) \, dW_T(s) \\
&= 2 \int\limits_{x_0}^{\xi_T(t)} g_0\left(G_T(u)\right) G'_T(u) \, du - 2 \int_0^t g_0\left(\zeta_T(s)\right) G'_T\left(\xi_T(s)\right) \, dW_T(s) \\
&\quad + 2 \int\limits_{x_0}^{\xi_T(t)} \widehat{q}_T(u) \, du - 2 \int_0^t \widehat{q}_T\left(\xi_T(s)\right) \, dW_T(s) \\
&= 2 \int\limits_{G_T(x_0)}^{\zeta_T(t)} g_0(u) \, du - 2 \int_0^t g_0\left(\zeta_T(s)\right) \, d\eta_T(s) + \gamma_T^{(1)}(t) - \gamma_T^{(2)}(t),
\end{aligned}
$$

where

$$\gamma_T^{(1)}(t) = \int\limits_{x_0}^{\xi_T(t)} \widehat{q}_T(u) \, du, \qquad \gamma_T^{(2)}(t) = \int\limits_0^t \widehat{q}_T\left(\xi_T(s)\right) \, dW_T(s),$$

$$\widehat{q}_T(x) = 2 \left(f'_T(x) \int\limits_0^x \frac{g_T(v)}{f'_T(v)} \, dv - g_0\left(G_T(x)\right) G'_T(x) \right).$$

Denote $P_{NT} = P \left\{ \sup\limits_{0 \le t \le L} |\xi_T(t)| > N \right\}$. It is clear that for any constants $\varepsilon > 0$, $N > 0$, and $L > 0$, we have the inequalities

$$P \left\{ \sup\limits_{0 \le t \le L} \left| \gamma_T^{(1)}(t) \right| > \varepsilon \right\} \le P_{NT} + \frac{2}{\varepsilon} E \sup\limits_{0 \le t \le L} \left| \int\limits_{x_0}^{\xi_T(t)} \widehat{q}_T(u) \, du \right| \chi_{\{|\xi_T(t)| \le N\}}$$

$$\le P_{NT} + \frac{2}{\varepsilon} \int\limits_{-N}^N |\widehat{q}_T(u)| \, du$$

and

$$P\left\{\sup_{0\leq t\leq L}\left|\gamma_T^{(2)}(t)\right| > \varepsilon\right\} \leq P_{NT} + \frac{4}{\varepsilon^2}\mathsf{E}\sup_{0\leq t\leq L}\left|\int_0^t \widehat{q}_T\left(\xi_T(s)\right)\chi_{\{|\xi_T(s))\leq N|\}}\,dW_T(s)\right|^2$$

$$\leq P_{NT} + \frac{16}{\varepsilon^2}\mathsf{E}\int_0^L \widehat{q}_T^2\left(\xi_T(s)\right)\chi_{\{|\xi_T(s))\leq N|\}}\,ds. \tag{5.23}$$

Using assumption (A_2) we obtain that

$$\lim_{N\to+\infty}\limsup_{T\to+\infty} P_{NT} = 0.$$

Thus, according to the conditions of Theorem 5.4, we get the convergence

$$\sup_{0\leq t\leq L}\left|\gamma_T^{(k)}(t)\right| \xrightarrow{\mathsf{P}} 0, \quad \text{for } k = 1, \tag{5.24}$$

as $T \to +\infty$.

Let us show that the last convergence holds with $k = 2$. Consider the function

$$\widetilde{\Phi}_T(x) = 2\int_0^x f_T'(u)\left(\int_0^u \frac{\widehat{q}_T^2(v)\chi_{|v|\leq N}}{f_T'(v)}\,dv\right)du.$$

We can apply the Itô formula to the process $\Phi_T(\xi_T(t))$, where $\xi_T(t)$ is a solution to Eq. (5.1).

Furthermore, the equality

$$\widetilde{\Phi}_T'(x)a_T(x) + \frac{1}{2}\widetilde{\Phi}_T''(x) = \widehat{q}_T^2(x)\chi_{|x|\leq N}$$

holds a.e. with respect to the Lebesgue measure. Using the latter equality and the properties of stochastic integrals, we conclude that

$$\mathsf{E}\int_0^L \widehat{q}_T^2\left(\xi_T(s)\right)\chi_{|(\xi_T(s))|\leq N}\,ds = \mathsf{E}\left[\widetilde{\Phi}_T(\xi_T(L)) - \widetilde{\Phi}_T(x_0)\right].$$

Using the Hölder inequality, we obtain the following bounds:

$$\left| \tilde{\Phi}_T(x) - \tilde{\Phi}_T(x_0) \right| \leq 2 \left| \int\limits_{x_0}^{x} f_T'(u) \left| \int\limits_{0}^{u} \frac{1}{|f_T'(v)|^q} \, dv \right|^{\frac{1}{q}} \left| \int\limits_{0}^{u} \left[\hat{q}_T^2(v) \right]^p \chi_{|v| \leq N} \, dv \right|^{\frac{1}{p}} du \right|$$

$$\leq 2 \left| \int\limits_{-N}^{N} |\hat{q}_T(v)|^{2p} \, dv \right|^{\frac{1}{p}} \left| \int\limits_{0}^{x} f_T'(u) \left| \int\limits_{0}^{u} \frac{1}{|f_T'(v)|^q} \, dv \right|^{\frac{1}{q}} du \right|,$$

where $p > 1$ and $q > 1$ are arbitrary constants with $\frac{1}{p} + \frac{1}{q} = 1$.

Taking into account assumption (A_3) for a certain $q = 1 + \delta$, we conclude that

$$\left| \mathsf{E} \left[\tilde{\Phi}_T(\xi_T(L)) - \tilde{\Phi}_T(x_0) \right] \right| \leq 2 \left(\int\limits_{-N}^{N} |\hat{q}_T(v)|^{2p} \, dv \right)^{\frac{1}{p}} C \left[1 + \mathsf{E} |\zeta_T(L)|^m \right],$$

where the constants $C > 0$ and $m \geq 0$ do not depend on T.

According to assumption (A_5) and to Lebesgue's dominated convergence theorem, we have the convergence $\int\limits_{-N}^{N} |\hat{q}_T(v)|^{2p} \, dv \to 0$, as $T \to +\infty$, for arbitrary $N > 0$. Hence, taking into account the inequality $\mathsf{E} |\zeta_T(L)|^m \leq C_L$ we obtain the convergence

$$\mathsf{E} \left[\tilde{\Phi}_T(\xi_T(L)) - \tilde{\Phi}_T(x_0) \right] \to 0,$$

as $T \to +\infty$, and consequently

$$\mathsf{E} \int\limits_{0}^{L} \hat{q}_T^2 \left(\xi_T(s) \right) \chi_{|(\xi_T(s))| \leq N} \, ds \to 0,$$

as $T \to +\infty$.

The latter convergence and inequality (5.23) imply the convergence (5.24) for $k = 2$. The same arguments as we used establishing (5.19) yield that

$$\tilde{\beta}_{T_n}^{(1)}(t) = 2 \int\limits_{G_{T_n}(x_0)}^{\tilde{\zeta}_{T_n}(t)} g_0(u) \, du - 2 \int\limits_{0}^{t} g_0 \left(\tilde{\zeta}_{T_n}(s) \right) d\tilde{\eta}_{T_n}(s) + \tilde{\gamma}_{T_n}^{(1)}(t) - \tilde{\gamma}_{T_n}^{(2)}(t),$$

$$(5.25)$$

where

$$\sup_{0 \le t \le L} \left| \widetilde{\zeta}_{T_n}(t) - \widetilde{\zeta}(t) \right| \overset{\widetilde{\mathrm{P}}}{\to} 0, \qquad \sup_{0 \le t \le L} \left| \widetilde{\eta}_{T_n}(t) - \widetilde{\eta}(t) \right| \overset{\widetilde{\mathrm{P}}}{\to} 0,$$

$$\sup_{0 \le t \le L} \left| \widetilde{\gamma}_{T_n}^{(k)}(t) \right| \overset{\widetilde{\mathrm{P}}}{\to} 0, \quad k = 1, 2$$

as $T_n \to +\infty$ for all $L > 0$. According to (5.20), the process $\left(\widetilde{\zeta}(t), \, \widehat{W}(t) \right)$ satisfies Eq. (5.14), and $\widetilde{\eta}(t)$ is defined in (5.21).

By Lemma 5.5 the finite-dimensional distributions of the stochastic process $\beta_{T_n}^{(1)}(t)$ coincide with those of the process $\widetilde{\beta}_{T_n}^{(1)}(t)$. Using Lemma 5.4, we can pass to the limit, as $T_n \to +\infty$, in (5.25) and obtain

$$\sup_{0 \le t \le L} \left| \widetilde{\beta}_{T_n}^{(1)}(t) - \widetilde{\beta}^{(1)}(t) \right| \overset{\widetilde{\mathrm{P}}}{\to} 0, \tag{5.26}$$

as $T_n \to +\infty$, where

$$\widetilde{\beta}^{(1)}(t) = 2 \int_{y_0}^{\widetilde{\zeta}(t)} g_0(u)\, du - 2 \int_0^t g_0\left(\widetilde{\zeta}(s) \right) d\widetilde{\eta}(s)$$

$$= 2 \left[\int_{y_0}^{\widetilde{\zeta}(t)} g_0(u)\, du - \int_0^t g_0\left(\widetilde{\zeta}(s) \right) d\widetilde{\zeta}(s) + \int_0^t g_0\left(\widetilde{\zeta}(s) \right) a_0\left(\widetilde{\zeta}(s) \right) ds \right],$$

and $\widetilde{\zeta}$ is a solution of Eq. (5.14). Therefore, we have that Theorem 5.4 holds for the process $\beta_{T_n}^{(1)}$, as $T_n \to +\infty$. Since the subsequence $T_n \to +\infty$ is arbitrary and since a solution ζ of Eq. (5.14) is weakly unique, the proof of Theorem 5.4 is complete.

$\square$

Theorem 5.5 *Let ξ_T be a solution of Eq. (5.1) belonging to the class $K\,(G_T)$, and let the assumptions of Theorem 5.2 hold. Let for the coefficient $a_T(x)$ of Eq. (5.1) assumption (A_5) hold. Assume that, for measurable and locally bounded functions $g_T(x)$, there exist certain constants c_T, m_T and $C > 0$ such that $|c_T| \le C$, $0 \le m_T \le C$ and for arbitrary $N > 0$*

$$\left| \int_0^x \left[f_T'(u) \int_0^u \frac{g_T(v)}{f_T'(v)}\, dv - c_T \right] du \right| \to 0,$$

a.e., as $T \to +\infty$,

$$\left| f'_T(x) \int_0^x \frac{g_T(v)}{f'_T(v)} \, dv - c_T \right| \chi_{|x| \le N} \le C_N$$

and for the functions

$$\widehat{q}_T(x) = \left[f'_T(u) \int_0^u \frac{g_T(v)}{f'_T(v)} \, dv - c_T \right]^2 - m_T^2$$

assumption (A_5) holds. Then,

(1) *for the subsequences $T_n \to +\infty$ such that $\liminf\limits_{T_n \to +\infty} m_{T_n} > 0$, the stochastic processes*

$$\widehat{\beta}_{T_n}^{(1)}(t) = \frac{1}{2m_{T_n}} \int_0^t g_{T_n}\left(\xi_{T_n}(s)\right) ds$$

converge weakly, as $T_n \to +\infty$, to the Wiener process W;

(2) *for the subsequences $T_n \to +\infty$ such that $\lim\limits_{T_n \to +\infty} m_{T_n} = 0$, the stochastic processes*

$$\beta_{T_n}^{(1)}(t) = \int_0^t g_{T_n}\left(\xi_{T_n}(s)\right) ds$$

converge weakly, as $T_n \to +\infty$, to zero.

Proof We apply the Itô formula to the process $\Phi_T(\xi_T(t))$, where ξ_T is a solution to Eq. (5.1) and the function Φ_T has the form

$$\Phi_T(x) = 2 \int_0^x f'_T(u) \left(\int_0^u \frac{g_T(v)}{f'_T(v)} \, dv \right) du.$$

As a result, we get the representation

$$\beta_T^{(1)}(t) = 2c_T \int_0^t a_T(\xi_T(s)) \, ds + \alpha_T(t) + \eta_T^{(1)}(t),$$

where

$$\beta_T^{(1)}(t) = \int_0^t g_T(\xi_T(s))\,ds,$$

$$\alpha_T(t) = 2 \int_{x_0}^{\xi_T(t)} \left[f_T'(u) \int_0^u \frac{g_T(v)}{f_T'(v)}\,dv - c_T \right] du,$$

$$\eta_T^{(1)}(t) = -\int_0^t \left[\Phi_T'(\xi_T(s)) - 2c_T \right] dW_T(s).$$

Let condition (A_5) hold for the functions $a_T(x)$. Then, according to Lemma 5.2, for arbitrary $L > 0$, we conclude

$$\sup_{0 \le t \le L} \left| \int_0^t a_T(\xi_T(s))\,ds \right| \overset{P}{\to} 0,$$

as $T \to +\infty$. We have the obvious inequalities

$$P\left\{ \sup_{0 \le t \le L} |\alpha_T(t)| > \varepsilon \right\} \le P_{NT} + \frac{2}{\varepsilon} E \sup_{0 \le t \le L} \left| \int_{x_0}^{\xi_T(t)} \left[\Phi_T'(u) - 2c_T \right] du \right| \chi_{|\xi_T(t)| \le N}$$

$$\le P_{NT} + \frac{4}{\varepsilon} \int_{-N}^{N} \left| f_T'(u) \int_0^u \frac{g_T(v)}{f_T'(v)}\,dv - c_T \right| du$$

for arbitrary $N > 0$, $L > 0$, and $\varepsilon > 0$, where $P_{NT} = P\left\{ \sup_{0 \le t \le L} |\xi_T(t)| > N \right\}$.
Taking into account that, according to assumption (A_2), $\lim_{N \to +\infty} \limsup_{T \to +\infty} P_{NT} = 0$, we get

$$\sup_{0 \le t \le L} |\alpha_T(t)| \overset{P}{\to} 0,$$

as $T \to +\infty$.
So we have

$$\sup_{0 \le t \le L} \left| \beta_T^{(1)}(t) - \eta_T^{(1)}(t) \right| \overset{P}{\to} 0,$$

as $T \to +\infty$.

It is clear that $\eta_T^{(1)}(t)$ is a continuous martingale with the quadratic characteristics

$$\langle \eta_T^{(1)} \rangle (t) = 4m_T^2 t + \int_0^t q_T(\xi_T(s))\, ds,$$

where

$$q_T(x) = 4 \left[f_T'(x) \int_0^x \frac{g_T(v)}{f_T'(v)}\, dv - c_T \right]^2 - 4m_T^2.$$

Condition (A_5) holds for the function $q_T(x)$, therefore, by Lemma 5.2

$$\sup_{0 \le t \le L} \left| \langle \eta_T^{(1)} \rangle (t) - 4m_T^2 t \right| \overset{P}{\to} 0,$$

as $T \to +\infty$, for any $L > 0$.

Next we use the random time change, i.e., $\eta_T^{(1)}(t) = W_T^* \left(\langle \eta_T^{(1)} \rangle (t) \right)$, where W_T^* is a Wiener process. Similarly to the proof of the relation (4.19) we have

$$\sup_{0 \le t \le L} \left| \beta_T^{(1)}(t) - W_T^* \left(4m_T^2 t \right) \right| \overset{P}{\to} 0, \tag{5.27}$$

as $T \to +\infty$.

Let $T_n \to +\infty$ be a subsequence such that $\liminf_{T_n \to +\infty} m_{T_n} > 0$. The process

$$\frac{W_{T_n}^* \left(4m_{T_n}^2 t \right)}{2m_{T_n}}$$

is a Wiener process for every T_n. Let us denote it as $W^*(t)$.

According to the convergence (5.27) and the equality

$$\sup_{0 \le t \le L} \left| \int_0^t g_{T_n}(\xi_{T_n}(s))\, ds - W_{T_n}^* \left(4m_{T_n}^2 t \right) \right| = 2m_{T_n} \sup_{0 \le t \le L} \left| \frac{1}{2m_{T_n}} \int_0^t g_{T_n}(\xi_{T_n}(s))\, ds - W^*(t) \right|$$

we have the convergence

$$\sup_{0 \le t \le L} \left| \frac{1}{2m_{T_n}} \int_0^t g_{T_n}(\xi_{T_n}(s))\, ds - W^*(t) \right| \to 0,$$

as $T_n \to +\infty$, whence the proof of statement (1) follows. Statement (2) follows from (5.27).   □

5.5   Asymptotic Behavior of Integral Functionals of Martingale Type

In this section we obtain the sufficient conditions for the weak convergence of some integral functionals of martingale type.

Theorem 5.6 *Let ξ_T be a solution of Eq. (5.1) belonging to the class $K(G_T)$ and let the assumptions of Theorem 5.2 hold. Assume that, for measurable and locally bounded functions g_T, there exists a measurable locally bounded function g_0 such that the function*

$$q_T(x) = \left[g_T(x) - g_0\left(G_T(x)\right) G_T'(x) \right]^2$$

satisfies assumption (A_5). Then the stochastic processes

$$\beta_T^{(2)}(t) = \int_0^t g_T(\xi_T(s))\, dW_T(s),$$

where $\xi_T(t)$ and $W_T(t)$ are related via Eq. (5.1), converge weakly, as $T \to +\infty$, to the process

$$\beta^{(2)}(t) = \int_0^t g_0(\zeta(s))\, \sigma_0(\zeta(s))\, d\widehat{W}(s),$$

where $\left(\zeta, \widehat{W}\right)$ is a solution to Eq. (5.14).

Proof It is clear that

$$\beta_T^{(2)}(t) = \int_0^t g_0(\zeta_T(s))\, d\eta_T(s) + \gamma_T(t), \tag{5.28}$$

where

$$\gamma_T(t) = \int_0^t q_T(\xi_T(s))\, dW_T(s), \qquad q_T(x) = g_T(x) - g_0\left(G_T(x)\right) G_T'(x).$$

The process $\gamma_T(t)$ for every $T > 0$ is continuous with probability 1 and a martingale with the quadratic characteristics

$$\langle \gamma_T \rangle(t) = \int\limits_0^t q_T^2(\xi_T(s))\, ds.$$

Since for the functions $q_T^2(x)$ assumption (A_5) holds, according to Lemma 5.2, we obtain that

$$\int\limits_0^L q_T^2\,(\xi_T(s))\, ds \xrightarrow{\text{P}} 0,$$

as $T \to +\infty$, for any constant $L > 0$.

For arbitrary constants $\varepsilon > 0$, $\delta > 0$, and $L > 0$ the following inequality holds (see Theorem A.7):

$$\mathsf{P}\left\{ \sup_{0 \le t \le L} |\gamma_T(t)| > \varepsilon \right\} \le \delta + \mathsf{P}\left\{ \int\limits_0^L q_T^2\,(\xi_T(s))\, ds > \varepsilon^2 \delta \right\}.$$

So, we have the convergence

$$\sup_{0 \le t \le L} |\gamma_T(t)| \xrightarrow{\text{P}} 0, \qquad (5.29)$$

as $T \to +\infty$, for any constant $L > 0$.

Similarly to the representation (5.19), for an arbitrary subsequence T_n we can get on certain probability space $(\widetilde{\Omega}, \widetilde{\mathfrak{F}}, \widetilde{\mathsf{P}})$, the equality

$$\widetilde{\beta}_{T_n}^{(2)}(t) = \int\limits_0^t g_0(\widetilde{\zeta}_{T_n}(s))\, d\widetilde{\eta}_{T_n}(s) + \widetilde{\gamma}_{T_n}(t),$$

where

$$\sup_{0 \le t \le L} \left|\widetilde{\zeta}_{T_n}(t) - \widetilde{\zeta}(t)\right| \xrightarrow{\widetilde{\text{P}}} 0, \qquad \sup_{0 \le t \le L} \left|\widetilde{\eta}_{T_n}(t) - \widetilde{\eta}(t)\right| \xrightarrow{\widetilde{\text{P}}} 0,$$

$$\sup_{0 \le t \le L} \left|\widetilde{\gamma}_{T_n}(t)\right| \xrightarrow{\widetilde{\text{P}}} 0,$$

as $T_n \to +\infty$, for any $L > 0$, the process $(\widetilde{\zeta}(t),\ \widehat{W}(t))$ satisfies Eq. (5.14), $\widetilde{\eta}(t)$ is defined in (5.21), and the processes $\widetilde{\beta}_{T_n}^{(2)}(t)$ and $\beta_{T_n}^{(2)}(t)$ are stochastically equivalent.

Similarly to the proof of the relation (5.26), we obtain

$$\sup_{0 \le t \le L} \left| \widetilde{\beta}_{T_n}^{(2)}(t) - \widetilde{\beta}^{(2)}(t) \right| \xrightarrow{\widetilde{P}} 0$$

as $T_n \to +\infty$, where

$$\widetilde{\beta}^{(2)}(t) = \int_0^t g_0\left(\widetilde{\zeta}(s)\right) d\widetilde{\eta}(s) = \int_0^t g_0\left(\widetilde{\zeta}(s)\right) d\widetilde{\zeta}(s) - \int_0^t g_0\left(\widetilde{\zeta}(s)\right) a_0\left(\widetilde{\zeta}(s)\right) ds.$$

Thus, the process $\widetilde{\beta}_{T_n}^{(2)}(t)$ converges weakly, as $T_n \to +\infty$, to the process $\widetilde{\beta}^{(2)}(t)$. Since the subsequence $T_n \to +\infty$ is arbitrary and since the processes $\widetilde{\beta}_{T_n}^{(2)}(t)$ and $\beta_{T_n}^{(2)}(t)$ are stochastically equivalent, the proof of Theorem 5.6 is complete.   □

Theorem 5.7 *Let ξ_T be a solution of Eq. (5.1) belonging to the class $K(G_T)$ and let the assumptions of Theorem 5.2 hold. Assume that, for measurable and locally bounded functions g_T there exist measurable locally bounded functions $\widehat{g}_T$ and g_0 such that for the functions*

$$q_T^{(1)}(x) = [g_T(x) - \widehat{g}_T(G_T(x))]^2,$$
$$q_T^{(2)}(x) = \widehat{g}_T^2(G_T(x)) - g_0^2(G_T(x))$$

assumption (A_5) holds, and the function $q_T^{(3)}(x) = |\widehat{g}_T(G_T(x))|$ satisfies assumptions (A_5) and (A_7).

Then the stochastic processes

$$\beta_T^{(2)}(t) = \int_0^t g_T(\xi_T(s)) dW_T(s),$$

where $\xi_T(t)$ and $W_T(t)$ are related via Eq. (5.1), converge weakly, as $T \to +\infty$, to the process $\beta^{(2)}(t) = W^*\left(\beta^{(1)}(t)\right)$, where

$$\beta^{(1)}(t) = \int_0^t g_0^2(\zeta(s)) ds,$$

here ζ is a solution to Eq. (5.14), $W^* = \{W^*(t),\ t \ge 0\}$ is a Wiener process and the processes $W^*(t)$ and $\beta^{(1)}(t)$ are independent.

Proof It is clear that

$$\beta_T^{(2)}(t) = \int_0^t \widehat{g}_T(\zeta_T(s)) \, dW_T(s) + \gamma_T(t), \qquad (5.30)$$

where

$$\gamma_T(t) = \int_0^t q_T(\xi_T(s)) \, dW_T(s), \quad q_T(x) = g_T(x) - \widehat{g}_T(G_T(x)).$$

Since for the functions $q_T^2(x)$ assumption (A_5) holds, we obtain relation (5.29) for the process $\gamma_T(t)$.

We can apply Skorokhod's convergent subsequence principle (see Theorem A.12) and obtain (5.19) on a probability space $(\widetilde{\Omega}, \widetilde{\mathfrak{F}}, \widetilde{\mathsf{P}})$ with additional representation

$$\widetilde{\beta}_{T_n}^{(2)}(t) = \int_0^t \widehat{g}_{T_n}(\widetilde{\zeta}_{T_n}(s)) \, d\widetilde{W}_{T_n}(s) + \widetilde{\gamma}_{T_n}(t).$$

Here

$$\sup_{0 \le t \le L} \left| \widetilde{\zeta}_{T_n}(t) - \widetilde{\zeta}(t) \right| \overset{\widetilde{\mathsf{P}}}{\to} 0, \quad \sup_{0 \le t \le L} \left| \widetilde{W}_{T_n}(t) - \widetilde{W}(t) \right| \overset{\widetilde{\mathsf{P}}}{\to} 0, \quad \sup_{0 \le t \le L} \left| \widetilde{\gamma}_{T_n}(t) \right| \overset{\widetilde{\mathsf{P}}}{\to} 0,$$

as $T_n \to +\infty$ for any $L > 0$, the process $\widetilde{\zeta}(t)$ is the solution to Eq. (5.14), and $\widetilde{W}(t)$ is a Wiener process.

According to Lemma 5.5, the processes $\widetilde{\beta}_{T_n}^{(2)}(t)$ and $\beta_{T_n}^{(2)}(t)$ are stochastically equivalent. Taking into account Lemma A.8, we use the random time change in stochastic integrals (see, e.g., [74]) and obtain that for any $t \ge 0$ with probability 1

$$\widetilde{\beta}_{T_n}^{(2)}(t) = W_{T_n}^*\left(\widetilde{\beta}_{T_n}^{(1)}(t)\right) + \widetilde{\gamma}_{T_n}(t), \qquad (5.31)$$

where $W_{T_n}^*(t)$ is a family of a Wiener processes,

$$\widetilde{\beta}_{T_n}^{(1)}(t) = \int_0^t \widehat{g}_{T_n}^2\left(\widetilde{\zeta}_{T_n}(s)\right) \, ds.$$

The functions $\widehat{g}_{T_n}^2 (G_T(x)) - g_0^2 (G_T(x))$ satisfy condition (A_5). The proof of Theorem 5.3 implies that

$$\sup_{0 \leq t \leq L} \left| \widetilde{\beta}_{T_n}^{(1)}(t) - \widetilde{\beta}^{(1)}(t) \right| \xrightarrow{\widetilde{P}} 0,$$

as $T_n \to +\infty$, where

$$\widetilde{\beta}^{(1)}(t) = \int_0^t g_0^2 \left(\widetilde{\zeta}(s) \right) \, ds.$$

It is clear that for any $L > 0$, $N > 0$, $\varepsilon > 0$, and $\delta > 0$ we have the inequality

$$\widetilde{P} \left\{ \sup_{0 \leq t \leq L} \left| W_{T_n}^* \left(\widetilde{\beta}_{T_n}^{(1)}(t) \right) - W_{T_n}^* \left(\widetilde{\beta}^{(1)}(t) \right) \right| > \varepsilon \right\}$$

$$\leq \widetilde{P} \left\{ \widetilde{\beta}_{T_n}^{(1)}(L) > N \right\} + \widetilde{P} \left\{ \widetilde{\beta}^{(1)}(L) > N \right\}$$

$$+ \widetilde{P} \left\{ \sup_{|t_1 - t_2| \leq \delta; \, t_i \leq N} |W_{T_n}^*(t_2) - W_{T_n}^*(t_1)| > \varepsilon \right\} + \widetilde{P} \left\{ \sup_{0 \leq t \leq L} \left| \widetilde{\beta}_{T_n}^{(1)}(t) - \widetilde{\beta}^{(1)}(t) \right| > \delta \right\}.$$

We have the analog of the convergence (5.12) for the Wiener process $W_{T_n}^*(t)$. Therefore, we obtain the convergence

$$\sup_{0 \leq t \leq L} \left| W_{T_n}^* \left(\widetilde{\beta}_{T_n}^{(1)}(t) \right) - W_{T_n}^* \left(\widetilde{\beta}^{(1)}(t) \right) \right| \xrightarrow{\widetilde{P}} 0,$$

as $T_n \to +\infty$. According to (5.31), we get

$$\sup_{0 \leq t \leq L} \left| \widetilde{\beta}_{T_n}^{(2)}(t) - W_{T_n}^* \left(\widetilde{\beta}^{(1)}(t) \right) \right| \xrightarrow{\widetilde{P}} 0, \tag{5.32}$$

as $T_n \to +\infty$. Using the properties of stochastic integrals, we obtain the inequality

$$\left| E W_{T_n}^*(t) \widetilde{W}_{T_n}(t) \right| = \left| E \int_0^{\tau_{T_n}(t)} \widehat{g}_{T_n} \left(\widetilde{\zeta}_{T_n}(s) \right) \, d\widetilde{W}_{T_n}(s) \widetilde{W}_{T_n}(t) \right| =$$

$$= \left| E \int_0^{t \wedge \tau_{T_n}(t)} \widehat{g}_{T_n} \left(\widetilde{\zeta}_{T_n}(s) \right) \, ds \right| \leq E \int_0^t \left| \widehat{g}_{T_n} \left(\widetilde{\zeta}_{T_n}(s) \right) \right| \, ds.$$

The conditions on the function $q_T^{(3)}(x)$, combined with Remark 5.2, imply the convergence

$$\mathsf{E} \int\limits_0^t \left| \widehat{g}_{T_n} \left(\widetilde{\zeta}_{T_n}(s) \right) \right| ds \to 0,$$

as $T_n \to +\infty$, for any $t > 0$. Thus,

$$\mathsf{E} W_{T_n}^*(t) \widetilde{W}_{T_n}(t) \to 0,$$

as $T_n \to +\infty$.

Since the processes $W_{T_n}^*(t)$ and $\widetilde{W}_{T_n}(t)$ are Wiener processes and they are asymptotically uncorrelated, we conclude that $W_{T_n}^*(t)$ does not depend asymptotically on $\widetilde{W}(t)$. It is clear that the process $\widetilde{\beta}^{(1)}(t)$ is completely determined by the process $\widetilde{\zeta}(s)$ for $s \leq t$.

Using the strong uniqueness of the solution $(\xi_T(t), W_T(t))$ to Eq. (5.1) we conclude that the processes $\widetilde{\zeta}(t)$ and $\widehat{W}(t)$ are measurable with respect to the σ-algebra $\sigma\left(\widetilde{W}(s), \ s \leq t\right)$, where $\widetilde{W}(t)$ is a Wiener process, which is the limit process for $\widetilde{W}_{T_n}(t)$. So, the process $W_{T_n}^*(t)$ does not depend asymptotically on the process $\widetilde{\beta}^{(1)}(t)$. The finite-dimensional distributions of the process $W_{T_n}^*(t)$ do not depend on T_n, thus, the limit process can be written as $W^*\left(\widetilde{\beta}^{(1)}(t)\right)$, where W^* is a Wiener process, and the processes W^* and $\widetilde{\beta}^{(1)}$ are independent. Taking into account (5.32), we have that

$$\sup_{0 \leq t \leq L} \left| \widetilde{\beta}_{T_n}^{(2)}(t) - W^* \left(\widetilde{\beta}^{(1)}(t) \right) \right| \xrightarrow{\widetilde{\mathsf{P}}} 0,$$

as $T_n \to +\infty$.

Therefore, the processes $\widetilde{\beta}_{T_n}^{(2)}(t)$ converge weakly, as $T_n \to +\infty$, to the process $W^*\left(\widetilde{\beta}^{(1)}(t)\right)$. It means that the statement of Theorem 5.7 is valid for the process $\beta_{T_n}^{(2)}$. Since the subsequence $T_n \to +\infty$ is arbitrary and since the finite-dimensional distributions of the process $W^*\left(\widetilde{\beta}^{(1)}(t)\right)$ are determined in a unique way, the proof of Theorem 5.7 is complete.   □

Theorem 5.8 *Let ξ_T be a solution of Eq. (5.1) belonging to the class $K(G_T)$, and let the assumptions of Theorem 5.2 hold. Assume that, for measurable and locally bounded functions g_T there exist measurable locally bounded functions $\widehat{g}_T$ and g_0 such that the function $q_T^{(1)}(x) = [g_T(x) - g_0(G_T(x))]^2$ satisfies assumption (A_5), the functions $q_T(x) = \widehat{g}_T^2(G_T(x))$ and $g_0(x)$ satisfy assumption (A_6), and the function $q_T^{(3)}(x) = |\widehat{g}_T(G_T(x))|$ satisfies assumptions (A_5) and (A_7).*

Then the stochastic processes

$$\beta_T^{(2)}(t) = \int\limits_0^t g_T(\xi_T(s)) \, dW_T(s),$$

where $\xi_T(t)$ and $W_T(t)$ are related via Eq. (5.1), converge weakly, as $T \to +\infty$, to the process $\beta^{(2)}(t) = W^* \left(\beta^{(1)}(t) \right)$, where

$$\beta^{(1)}(t) = 2 \left[\int\limits_{y_0}^{\zeta(t)} g_0(x) \, dx + \int\limits_0^t g_0(\zeta(s)) \, \sigma_0(\zeta(s)) \, d\widehat{W}(s) \right], \qquad (5.33)$$

here $\left(\zeta(t), \widehat{W}(t) \right)$ is a solution to Eq. (5.14), $W^* = \{W^*(t), \ t \ge 0\}$ is a Wiener process, and the processes W^* and $\beta^{(1)}$ are independent.

Proof The proof of Theorem 5.8 completely coincides with the proof of Theorem 5.7 with the only difference that the process $\beta^{(1)}$ has the form (5.33).   □

5.6   Weak Convergence of Mixed Functionals

In this section we obtain sufficient conditions for the weak convergence of some mixed functionals.

Theorem 5.9 *Let ξ_T be a solution of Eq. (5.1) belonging to the class $K(G_T)$, and let the assumptions of Theorem 5.2 hold. Assume that, for continuous functions F_T and a locally bounded measurable functions g_T, there exist a continuous function F_0 and locally bounded measurable function g_0 such that, for arbitrary $N > 0$*

$$\lim_{T \to +\infty} \sup_{|x| \le N} |F_T(x) - F_0(G_T(x))| = 0.$$

Also, let the function

$$q_T(x) = \left[g_T(x) - g_0(G_T(x)) \, G_T'(x) \right]^2$$

satisfy assumption (A_5). Then the stochastic processes

$$I_T(t) = F_T(\xi_T(t)) + \int\limits_0^t g_T(\xi_T(s)) \, dW_T(s),$$

where ξ_T and W_T are related via Eq. (5.1), converge weakly, as $T \to +\infty$, to the process

$$I_0(t) = F_0(\zeta(t)) + \int\limits_0^t g_0(\zeta(s))\,\sigma_0(\zeta(s))\,d\widehat{W}(s),$$

where $\left(\zeta(t),\ \widehat{W}(t)\right)$ is a solution of Eq. (5.14).

Proof It is clear that

$$I_T(t) = F_0(\zeta_T(t)) + \int\limits_0^t g_0(\zeta_T(s))\,d\eta_T(s) + \alpha_T(t) + \gamma_T(t),$$

where

$$\alpha_T(t) = F_T\,(\xi_T(t)) - F_0\,(\zeta_T(t))\,, \quad \eta_T(t) = \int\limits_0^t G_T'(\xi_T(s))\,dW_T(s),$$

$$\gamma_T(t) = \int\limits_0^t q_T\,(\xi_T(s))\,dW_T(s), \quad q_T(x) = g_T(x) - g_0\,(G_T(x))\,G_T'(x).$$

Denote as before, $P_{NT} = P\left\{\sup\limits_{0 \le t \le L} |\xi_T(t)| > N\right\}$. Since for any constants $\varepsilon > 0$, $N > 0$ and $L > 0$ the following inequality holds:

$$P\left\{\sup\limits_{0 \le t \le L} |F_T\,(\xi_T(t)) - F_0\,(G_T(\xi_T(t)))| > \varepsilon\right\}$$

$$\le P_{NT} + \frac{2}{\varepsilon}E\sup\limits_{0 \le t \le L} |F_T\,(\xi_T(t)) - F_0\,(G_T(\xi_T(t)))|\,\chi_{|\xi_T(t)| \le N}$$

$$\le P_{NT} + \frac{2}{\varepsilon}\sup\limits_{|x| \le N} |F_T\,(x) - F_0\,(G_T(x))|\,,$$

we can apply the conditions of Theorem 5.9 and condition (A_2) to get that

$$\sup\limits_{0 \le t \le L} |\alpha_T(t)| \xrightarrow{P} 0,$$

as $T \rightarrow +\infty$. The proof of the fact that for $\gamma_T(t)$ the convergence (5.29) holds is literally the same as the respective part of the proof of Theorem 5.6. Then, we can apply Skorokhod's convergent subsequence principle for the process $(\zeta_T(t), \eta_T(t), \alpha_T(t), \gamma_T(t))$ and, similarly to the representation (5.19), obtain equality for an arbitrary subsequence T_n in a certain probability space $(\widetilde{\Omega}, \widetilde{\mathfrak{F}}, \widetilde{\mathsf{P}})$

$$\widetilde{I}_{T_n}(t) = F_0(\widetilde{\zeta}_{T_n}(t)) + \int_0^t g_0(\widetilde{\zeta}_{T_n}(s)) \, d\widetilde{\eta}_{T_n}(s) + \widetilde{\alpha}_{T_n}(t) + \widetilde{\gamma}_{T_n}(t),$$

where for any $L > 0$

$$\sup_{0 \le t \le L} \left| \widetilde{\zeta}_{T_n}(t) - \widetilde{\zeta}(t) \right| \xrightarrow{\widetilde{\mathsf{P}}} 0, \qquad \sup_{0 \le t \le L} \left| \widetilde{\eta}_{T_n}(t) - \widetilde{\eta}(t) \right| \xrightarrow{\widetilde{\mathsf{P}}} 0,$$

$$\sup_{0 \le t \le L} \left| \widetilde{\alpha}_{T_n}(t) \right| \xrightarrow{\widetilde{\mathsf{P}}} 0, \qquad \sup_{0 \le t \le L} \left| \widetilde{\gamma}_{T_n}(t) \right| \xrightarrow{\widetilde{\mathsf{P}}} 0,$$

as $T_n \rightarrow +\infty$.

Since the function $F_0(x)$ is continuous, it is uniformly continuous in a closed region $|x| \le N$ for any $N > 0$. Denote

$$\widetilde{P}_{NT_n} = \mathsf{P} \left\{ \sup_{0 \le t \le L} \left| \widetilde{\zeta}_{T_n}(t) \right| > N \right\} \quad \text{and} \quad \widetilde{P}_N = \mathsf{P} \left\{ \sup_{0 \le t \le L} \left| \widetilde{\zeta}(t) \right| > N \right\}.$$

For any constants $\varepsilon > 0$, $N > 0$, and $L > 0$ there exists $\delta > 0$ such that the following inequalities hold:

$$\mathsf{P} \left\{ \sup_{0 \le t \le L} \left| F_0\left(\widetilde{\zeta}_{T_n}(t)\right) - F_0\left(\widetilde{\zeta}(t)\right) \right| > \varepsilon \right\} \le \widetilde{P}_{NT_n} + \widetilde{P}_N$$

$$+ \mathsf{P} \left\{ \left(\sup_{0 \le t \le L} \left| F_0\left(\widetilde{\zeta}_{T_n}(t)\right) - F_0\left(\widetilde{\zeta}(t)\right) \right| > \varepsilon \right) \cap \left(\sup_{0 \le t \le L} \left| \widetilde{\zeta}_{T_n}(t) \right| \le N \right) \right.$$

$$\left. \cap \left(\sup_{0 \le t \le L} \left| \widetilde{\zeta}(t) \right| \le N \right) \right\} \le \widetilde{P}_{NT_n} + \widetilde{P}_N$$

$$+ \mathsf{P} \left\{ \left(\sup_{0 \le t \le L} \left| \widetilde{\zeta}_{T_n}(t) - \widetilde{\zeta}(t) \right| > \delta \right) \cap \left(\sup_{0 \le t \le L} \left| \widetilde{\zeta}_{T_n}(t) \right| \le N \right) \cap \left(\sup_{0 \le t \le L} \left| \widetilde{\zeta}(t) \right| \le N \right) \right\}$$

$$\leq \widetilde{P}_{NT_n} + \widetilde{P}_N + \mathsf{P} \left\{ \sup_{0 \leq t \leq L} \left| \widetilde{\zeta}_{T_n}(t) - \widetilde{\zeta}(t) \right| > \delta \right\}.$$

Therefore,

$$\sup_{0 \leq t \leq L} \left| \widetilde{I}_{T_n}(t) - F_0\left(\widetilde{\zeta}(t)\right) - \int_0^t g_0\left(\widetilde{\zeta}_{T_n}(s)\right) d\widetilde{\eta}_{T_n}(s) \right| \overset{\widetilde{\mathsf{P}}}{\to} 0, \tag{5.34}$$

as $T_n \to +\infty$. Using Lemma 5.4 we can pass to the limit, as $T_n \to +\infty$, in the stochastic integral from (5.34) and obtain

$$\sup_{0 \leq t \leq L} \left| \widetilde{I}_{T_n}(t) - F_0\left(\widetilde{\zeta}(t)\right) - \int_0^t g_0\left(\widetilde{\zeta}(s)\right) d\widetilde{\eta}(s) \right| \overset{\widetilde{\mathsf{P}}}{\to} 0,$$

as $T_n \to +\infty$. It is an analog of convergence (5.26). From here we have that

$$\sup_{0 \leq t \leq L} \left| F_0\left(\zeta_{T_n}(t)\right) - F_0\left(\widetilde{\zeta}(t)\right) \right| \overset{\widetilde{\mathsf{P}}}{\to} 0,$$

as $T_n \to +\infty$.

To complete the proof of Theorem 5.9, we repeat the same arguments as in the proof of Theorem 5.6.   □

Theorem 5.10 *Let ξ_T be a solution of Eq. (5.1) belonging to the class $K\,(G_T)$ and let the assumptions of Theorem 5.2 hold. Assume that, for continuous functions F_T and a locally bounded measurable functions g_T, there exist a continuous function F_0 and locally bounded measurable functions $\widehat{g}_T$ and g_0 such that, for arbitrary $N > 0$*

$$\lim_{T \to +\infty} \sup_{|x| \leq N} |F_T(x) - F_0\,(G_T(x))| = 0,$$

and let for the function g_T, $\widehat{g}_T$ and g_0 the assumptions of Theorem 5.7 hold.
   Then the stochastic processes

$$I_T(t) = F_T(\xi_T(t)) + \int_0^t g_T(\xi_T(s))\,dW_T(s),$$

where ξ_T and W_T are related via Eq. (5.1), converge weakly, as $T \to +\infty$, to the process

$$\widehat{I}_0(t) = F_0(\zeta(t)) + W^*\left(\beta^{(1)}(t)\right),$$

where

$$\beta^{(1)}(t) = \int\limits_0^t g_0^2(\zeta(s))\,ds,$$

here ζ is a solution to Eq. (5.14), $W^* = \{W^*(t),\ t \geq 0\}$ is a Wiener process, and the processes W^* and $\beta^{(1)}$ are independent.

Proof It is clear that

$$I_T(t) = F_0(\zeta_T(t)) + \int\limits_0^t \widehat{g}_T(\zeta_T(s))\,dW_T(s) + \alpha_T(t) + \gamma_T(t),$$

where

$$\alpha_T(t) = F_T(\xi_T(t)) - F_0(\zeta_T(t)), \quad \gamma_T(t) = \int\limits_0^t q_T(\xi_T(s))\,dW_T(s),$$

$$q_T(x) = g_T(x) - \widehat{g}_T(G_T(x)).$$

Since the function $q_T^2(x)$ satisfies the assumptions of Lemma 5.2, we have

$$\sup_{0 \leq t \leq L} \int\limits_0^L q_T^2(\xi_T(s)) \xrightarrow{\mathrm{P}} 0,$$

as $T_n \to +\infty$, for any $L > 0$.

Thus, for the martingale $\gamma_T(t)$ we have the analog of the convergence (5.29). The proof of this fact is literally the same as in the proof of Theorem 5.6.

Then, taking into account the proof of Theorem 5.9 we can apply Skorokhod's convergent subsequence principle to the process

$$(\zeta_T(t),\ W_T(t),\ \alpha_T(t),\ \gamma_T(t))$$

and obtain equality for an arbitrary subsequence T_n in a certain probability space $(\widetilde{\Omega}, \widetilde{\mathfrak{F}}, \widetilde{\mathrm{P}})$

$$\widetilde{I}_{T_n}(t) = F_0(\widetilde{\zeta}_{T_n}(t)) + \int\limits_0^t \widehat{g}_{T_n}(\widetilde{\zeta}_{T_n}(s))\,d\widetilde{W}_{T_n}(s) + \widetilde{\alpha}_{T_n}(t) + \widetilde{\gamma}_{T_n}(t),$$

where for any $L > 0$

$$\sup_{0 \leq t \leq L} \left| \widetilde{\zeta}_{T_n}(t) - \widetilde{\zeta}(t) \right| \xrightarrow{\widetilde{P}} 0, \qquad \sup_{0 \leq t \leq L} \left| \widetilde{W}_{T_n}(t) - \widetilde{W}(t) \right| \xrightarrow{\widetilde{P}} 0,$$

$$\sup_{0 \leq t \leq L} \left| \widetilde{\alpha}_{T_n}(t) \right| \xrightarrow{\widetilde{P}} 0, \qquad \sup_{0 \leq t \leq L} \left| \widetilde{\gamma}_{T_n}(t) \right| \xrightarrow{\widetilde{P}} 0,$$

as $T_n \to +\infty$. Here $\widetilde{\zeta}$ is the solution to Eq. (5.14), $\widetilde{W}$ is a Wiener process.

Taking into account Lemma A.8, we change the time in the stochastic integral

$$\int_0^t \widehat{g}_{T_n}(\widetilde{\zeta}_{T_n}(s)) \, d\widetilde{W}_{T_n}(s)$$

and obtain that for any $t \geq 0$ with probability 1

$$\widetilde{I}_{T_n}(t) = F_0(\widetilde{\zeta}_{T_n}(t)) + W^*_{T_n} \left(\int_0^t \widehat{g}^2_{T_n}(\widetilde{\zeta}_{T_n}(s)) \, ds \right) + \widetilde{\alpha}_{T_n}(t) + \widetilde{\gamma}_{T_n}(t),$$

where $W^*_{T_n}(t)$ is the Wiener process for every T_n.

To complete the proof of Theorem 5.10 we use the equivalence of the processes $I_{T_n}(t)$ and $\widetilde{I}_{T_n}(t)$ and repeat the same arguments as in the proof of Theorem 5.7. □

Theorem 5.11 *Let ξ_T be a solution of Eq. (5.1) belonging to the class $K(G_T)$, and let the assumptions of Theorem 5.2 hold. Assume that, for continuous functions F_T and a locally bounded measurable functions g_T, there exist a continuous function F_0 and a locally bounded measurable functions $\widehat{g}_T$ and g_0 such that, for arbitrary $N > 0$*

$$\lim_{T \to +\infty} \sup_{|x| \leq N} |F_T(x) - F_0(G_T(x))| = 0.$$

Also, let for the functions $g_T(x)$, $\widehat{g}_T(x)$, and $g_0(x)$ the assumptions of Theorem 5.8 hold. Then the stochastic processes

$$I_T(t) = F_T(\xi_T(t)) + \int_0^t g_T(\xi_T(s)) \, dW_T(s),$$

where ξ_T and W_T are related via Eq. (5.1), converge weakly, as $T \to +\infty$, to the process

$$\widehat{I}_0(t) = F_0(\zeta(t)) + W^* \left(\beta^{(1)}(t) \right),$$

where

$$\beta^{(1)}(t) = 2 \left[\int\limits_{y_0}^{\zeta(t)} g_0(x)\, dx + \int\limits_{0}^{t} g_0(\zeta(s))\sigma_0\,(\zeta(s))\, d\widehat{W}(s) \right], \qquad (5.35)$$

here $\left(\zeta, \widehat{W}\right)$ is a solution of Eq. (5.14), and the processes W^* and $\widehat{W}$ are independent Wiener processes.

Proof The proof of Theorem 5.11 completely coincides with the proof of Theorem 5.10 with the only difference that the process $\beta^{(1)}$ has form (5.35). □

5.7 Examples

Denote by b_T the family of constants such that $b_T > 1$ and $b_T \uparrow +\infty$, as $T \to +\infty$.
Consider the following examples of the coefficients $a_T(x)$ of Eq. (5.1).

Example 5.1 Let

$$a_T(x) = \frac{b_T}{1 + x^2 b_T^2} + \frac{b_T}{1 + (x-1)^2 b_T^2} - \frac{1}{2}\frac{x}{1 + x^2},$$

and let the initial condition in Eq. (5.1) be $x_0 = 0$.
Taking into account the equality

$$f_T'(x) = \exp\{-2\arctan b_T x\} \cdot \exp\{-2\,[\arctan b_T(x-1) + \arctan b_T]\} \cdot \sqrt{1 + x^2},$$

we conclude that the solution $\xi_T(t)$ of Eq. (5.1) belongs to the class $K(G_T)$ for

$$G_T(x) = f_T(x) = \int\limits_0^x \exp\left\{-2\int\limits_0^u a_T(v)\, dv\right\} du.$$

In this case the conditions of Theorem 5.2 are fulfilled with $a_0(x) \equiv 0$ and $\sigma_0(x) = \overline{\sigma}\,(\varphi(x))\,\sqrt{1 + \varphi^2(x)}$, where

$$\overline{\sigma}(x) = \begin{cases} e^{\pi} & \text{for } x < 0, \\ e^{-\pi} & \text{for } 0 \le x < 1, \\ e^{-3\pi} & \text{for } x \ge 1, \end{cases}$$

and $\varphi(x)$ is the inverse function to $l(x) = \int\limits_0^x \overline{\sigma}(u)\sqrt{1 + u^2}\, du$.

Therefore, according to Theorem 5.2, the stochastic processes $\zeta_T = f_T(\xi_T(t))$ converge weakly, as $T \to +\infty$, to the solution ζ of the equation $\zeta(t) = \int_0^t \sigma_0(\zeta(s)) \, d\widehat{W}(s)$.

Example 5.2 Let $a_T(x) \equiv 0$ and $x_0 = 0$. Consider the family of functions

$$g_T(x) = b_T \left(\sin\left[(x - x_1) \, b_T \right] + \cos\left[(x - x_2) \, b_T \right] \right).$$

It is clear that the solution of Eq. (5.1) has the form $\xi_T(t) = W_T(t)$. In this case Eq. (5.1) belongs to the class $K(G_T)$ for $G_T(x) = x$, $f_T'(x) = 1$, $a_0(x) = 0$, $\sigma_0(x) = 1$.

The conditions of Theorem 5.5 hold for

$$c_T = \cos(x_1 b_T) + \sin(x_2 b_T) \quad \text{and} \quad m_T^2 = 1 + \sin\left((x_2 - x_1) \, b_T \right).$$

The stochastic processes

$$\widehat{\beta}_{T_n}^{(1)}(t) = \frac{1}{2m_{T_n}} \int_0^t g_{T_n}\left(W_{T_n}(s) \right) ds$$

for the subsequences $T_n \to +\infty$ such that $\liminf\limits_{T_n \to +\infty} m_{T_n} > 0$, converge weakly, as $T_n \to +\infty$, to a Wiener process W.

The proof of Theorem 5.5 implies that for the subsequences $T_n \to +\infty$ such that $m_{T_n} \to 0$ the stochastic processes

$$\beta_{T_n}^{(1)}(t) = \int_0^t g_{T_n}\left(W_{T_n}(s) \right) ds$$

converge weakly, as $n \to +\infty$, to the process $\beta(t) \equiv 0$.

In particular, the functional of the form

$$\beta_T^{(1)}(t) = \int_0^t g_T\left(W_T(s) \right) ds,$$

for $x_1 = x_2$ converges weakly, as $T \to +\infty$, to the process $\beta(t) = 2W^*(t)$, where W^* is a Wiener process. If $x_1 < x_2$, then the stochastic processes $\beta_{T_n}^{(1)}(t)$ for $T_n = \frac{1}{x_2 - x_1}\left[\frac{2\pi}{3} + 2n\pi \right]$ converge weakly, as $n \to +\infty$, to the process $\beta(t) \equiv 0$.

Example 5.3 Let $a_T(x) \equiv 0$ in Eq. (5.1). Then $\xi_T(t) = x_0 + W_T(t)$ is the solution of Eq. (5.1) belonging to the class $K(G_T)$ for $G_T(x) = x$. In this case, the conditions of Theorem 5.2 are fulfilled with $a_0(x) = 0$ and $\sigma_0(x) = 1$. The assumptions of Theorem 5.5 with $c_T = \cos b_T x_1$ and $m_T^2 = \frac{1}{2}$ hold for $g_T(x) = b_T \sin (b_T (x - x_1))$. According to Theorem 5.2, the stochastic processes $\xi_T(t)$ converge weakly, as $T \to +\infty$, to the process $\zeta(t) = x_0 + \widehat{W}(t)$, where $\widehat{W}$ is a Wiener process. According to Theorem 5.5, the stochastic processes

$$\beta_T^{(1)}(t) = \int_0^t b_T \sin (b_T \xi_T(s)) \, ds$$

converge weakly, as $T \to +\infty$, to the process $\sqrt{2}\widehat{W}(t)$.

Example 5.4 Let in Eq. (5.1) $a_T(x) \equiv 0$. Consider

$$g_T(x) = \sqrt{\frac{b_T}{1 + b_T^2 x^2}}.$$

It is clear that $g_T^2(x)$ is a δ-shaped family at the point $x = 0$ with weight π. The assumptions of Theorem 5.8 hold with $G_T(x) = x$ and $g_0(x) = \frac{\pi}{2} \operatorname{sign} x$.

Thus, Theorem 5.8 implies that the stochastic processes

$$\beta_T^{(2)}(t) = \int_0^t \sqrt{\frac{b_T}{1 + b_T^2 W_T^2(s)}} \, dW_T(s)$$

converge weakly to the process $\beta^{(2)}(t) = W^* (\beta^{(1)}(t))$, as $T \to +\infty$, where

$$\beta^{(1)}(t) = 2 \left[\int_{x_0}^{\zeta(t)} \frac{\pi}{2} \operatorname{sign} x \, dx - \int_0^t \frac{\pi}{2} \operatorname{sign} \zeta(s) \, dW(s) \right],$$

$\zeta(t) = x_0 + W(t)$, and the processes W^* and W are independent.

Thus, $\beta^{(2)}(t) = W^* (\pi L_W^{x_0}(t))$, where

$$L_W^{x_0}(t) = |x_0 + W(t)| - |x_0| - \int_0^t \operatorname{sign} (x_0 + W(s)) \, dW(s)$$

is the local time on the interval $[0, t]$ of a Wiener process W at the point x_0.

Example 5.5 Consider Eq. (5.1), where $a_T(x) = b_T \left[1 + (b_T x - 1)^2 \right]^{-1}$ is a δ-shaped family at the point $x = 0$ with weight π.

We are going to show that Eq. (5.1) belongs to the class $K(G_T)$ for

$$G_T(x) = f_T(x) = \int_0^x \exp\left\{-2\int_0^u a_T(v)\,dv\right\}\,du.$$

Indeed,

$$f_T'(x) = \exp\left\{-2\int_0^x a_T(v)\,dv\right\} = \exp\left\{-2arctg(b_T v - 1)\big|_0^x\right\}$$

$$= \exp\left\{-2\left[arctg(b_T x - 1) + arctg 1\right]\right\} \to \sigma_0(x) = \begin{cases} e^{-\frac{3}{2}\pi}, & x > 0, \\ e^{\frac{\pi}{2}}, & x < 0, \end{cases} \quad \text{as } T \to +\infty.$$

Since $f_T'(x)a_T(x) + \frac{1}{2}f_T''(x) = 0$, we have

$$\left[G_T'(x)a_T(x) + \frac{1}{2}G_T''(x)\right]^2 + \left[G_T'(x)\right]^2 = \left[f_T'(x)\right]^2 \leq C \leq C\left[1 + |G_T(x)|^2\right].$$

We derive $|G_T(x)| \geq C|x|^\alpha$ for all $x \in \mathbb{R}$ with $C = \delta_0$ and $\alpha = 1$ from the inequalities $0 < \delta_0 \leq G_T'(x) = f_T'(x) \leq C_0$. In addition,

$$\left|\int_0^x f_T'(u)\left(\int_0^u \frac{\chi_B(G_T(v))}{f_T'(v)}\,dv\right)du\right| \leq \frac{C_0}{\delta_0}\left|\int_0^x \int_0^u \chi_B(G_T(v))\,dv\,du\right|$$

$$\leq C_1\lambda(B)|x| \leq \psi(\lambda(B))\left[1 + |x|^{m_2}\right].$$

Therefore, condition (A_4) holds for the case of $\psi(|x|) = C_1|x|$ and $m_2 = 1$.

Let us check the assumptions of Theorem 5.2:

$$q_T^{(1)}(x) = G_T'(x)a_T(x) + \frac{1}{2}G_T''(x) - a_0(G_T(x)) \equiv 0,$$

$$q_T^{(2)}(x) = \left[G_T'(x)\right]^2 - \sigma_0^2(G_T(x)) = \begin{cases} \left[G_T'(x)\right]^2 - e^{-3\pi} \to 0, & x > 0, \\ \left[G_T'(x)\right]^2 - e^{\pi} \to 0, & x < 0, \end{cases} \quad \text{as } T \to +\infty.$$

and

$$\sup_{|x| \leq N} f_T'(x)\left|\int_0^x \frac{q_T^{(2)}(v)}{f_T'(v)}\,dv\right| \leq \frac{C_0}{\delta_0}\int_{-N}^N \left|q_T^{(2)}(v)\right|\,dv \to 0, \quad \text{when } T \to +\infty.$$

Therefore, the conditions of Theorem 5.2 hold for $a_0(x) \equiv 0$ and $\sigma_0(x) = e^{-\frac{3}{2}\pi}$, if $x > 0$, $\sigma_0(x) = e^{\frac{\pi}{2}}$, if $x \leq 0$, $y_0 = x_0\sigma_0(x_0)$. Thus, the stochastic process $\zeta_T(t) =$

$G_T(\xi_T(t))$ converges weakly, as $T \to +\infty$, to the solution ζ of the following Itô's equation:

$$\zeta(t) = x_0 \sigma_0(x_0) + \int_0^t \sigma_0\left(\zeta(s)\right) d\widehat{W}(s).$$

The assumptions of Theorem 5.8 hold for the functions

$$g_T(x) = \cos(b_T x)$$

if $\widehat{g}_T(x) = g_T\left(G_T^{-1}(x)\right)$, where $G_T^{-1}(x)$ are the inverse functions to $G_T(x)$ and $g_0(x) \equiv \frac{1}{2}$. In this case $\beta^{(1)}(t) = \frac{1}{2}t$. Hence, the stochastic processes

$$\beta_T^{(2)}(t) = \int_0^t \cos(b_T \xi_T(s)) \, dW_T(s)$$

converge weakly, as $T \to +\infty$, to the process $\frac{1}{\sqrt{2}} W^*(t)$, where W^* is a Wiener process.

Example 5.6 Let in Eq. (5.1)

$$a_T(x) = -\frac{1}{4} \frac{b_T^2 x}{1 + b_T^2 x^2}.$$

In this case, Eq. (5.1) belongs to the class $K(G_T)$ for $G_T(x) = x^2$. The assumptions of Theorem 5.2 hold for $a_0(x) = \frac{1}{2}$, $\sigma_0(x) = 2\sqrt{|x|}$, and $y_0 = x_0^2$. Therefore, the stochastic processes $\zeta_T(t) = \xi_T^2(t)$ converge weakly, as $T \to +\infty$, to the solution ζ of equation

$$\zeta(t) = x_0^2 + \frac{1}{2}t + 2 \int_0^t \sqrt{\zeta(s)} \, d\widehat{W}(s). \tag{5.36}$$

The assumptions of Theorem 5.8 hold for the functions

$$g_T(x) = \frac{\sqrt[4]{b_T}}{\sqrt{\ln b_T}} \frac{\cos(b_T x)}{\sqrt[8]{1 + b_T^2 x^2}}$$

if

$$g_0(x) = \frac{1}{4} \frac{1}{\sqrt[4]{|x|}}, \quad \widehat{g}_T(x) = \frac{\sqrt[4]{b_T}}{\sqrt{\ln b_T}} \frac{\cos\left(b_T\sqrt{|x|}\right)}{\sqrt[8]{1 + b_T^2 x^2}}.$$

In this case, $\widehat{g}_T(x^2) = g_T(x)$. According to Theorem 5.8, the stochastic processes

$$\beta_T^{(2)}(t) = \frac{\sqrt[4]{b_T}}{\sqrt{\ln b_T}} \int_0^t \frac{\cos\left(b_T \xi_T(s)\right)}{\sqrt[8]{1 + b_T^2 \xi_T^2(s)}} \, dW_T(s)$$

converge weakly, as $T \to +\infty$, to the process $W^*\left(\beta^{(1)}(t)\right)$, where

$$\beta^{(1)}(t) = \frac{2}{3}\left[\zeta^{\frac{3}{4}}(t) - |x_0|^{\frac{3}{2}}\right] - \int_0^t \sqrt[4]{\zeta(s)} \, d\widehat{W}(s),$$

$(\zeta, \widehat{W})$ is a solution of Eq. (5.36), and W^* is a Wiener process such that W^* and $\beta^{(1)}$ are independent.

Example 5.7 Let $a_T(x) = b_T \chi_{\left[0, \frac{\lambda}{b_T}\right]}(x)$ and $\lambda > 0$. If

$$G_T(x) = f_T(x) = \int_0^x \exp\left\{-2\int_0^u a_T(v)\,dv\right\} du,$$

then Eq. (5.1) belongs to the class $K(G_T)$. The assumptions of Theorem 5.2 hold for $a_0(x) = 0$,

$$\sigma_0(x) = \begin{cases} e^{-2\lambda}, & x > 0, \\ 1, & x \le 0, \end{cases}$$

and $y_0 = x_0\sigma_0(x_0)$. Thus, the stochastic processes $\zeta_T(t) = G_T(\xi_T(t))$ converge weakly, as $T \to +\infty$, to the solution ζ of the following Itô's equation:

$$\zeta(t) = x_0\sigma_0(x_0) + \int_0^t \sigma_0(\zeta(s)) \, d\widehat{W}(s). \tag{5.37}$$

The assumptions of Theorem 5.8 with

$$g_0(x) = \frac{\pi}{2} \frac{\operatorname{sign} x}{\sigma_0(x)} \quad \text{and} \quad \widehat{g}_T(x) = \left(\frac{b_T}{1 + b_T^2 \left[G_T^{-1}(x) \right]^2} \right)^{\frac{1}{2}}$$

hold for the functions

$$g_T(x) = \left(\frac{b_T}{1 + b_T^2 x^2} \right)^{\frac{1}{2}},$$

where $G_T^{-1}(x)$ denotes the inverse function to $G_T(x)$.

According to Theorem 5.8, the stochastic processes

$$\beta_T^{(2)}(t) = \int_0^t \sqrt{\frac{b_T}{1 + b_T^2 \xi_T^2(s)}} \, dW_T(s) \tag{5.38}$$

converge weakly, as $T \to +\infty$, to the process $W^* \left(\beta^{(1)}(t) \right)$, where

$$\beta^{(1)}(t) = \pi \left[\int_{x_0 \sigma_0(x_0)}^{\zeta(t)} \frac{\operatorname{sign} v}{\sigma_0(v)} \, dv - \int_0^t \operatorname{sign} \zeta(s) \, d\widehat{W}(s) \right],$$

$\left(\zeta, \, \widehat{W} \right)$ is a solution of Eq. (5.37), and W^* is a Wiener process, W^* and $\beta^{(1)}$ are independent.

Remark 5.1 The classes $K(G_T)$, related to Eq. (5.1), are not defined uniquely. In particular, if $a_T(x)$ in Eq. (5.1) are the same as in Example 5.7, then Eq. (5.1) belongs to the class $K(G_T)$ with $G_T(x) = x^2$. In addition, according to Theorem 5.2 with $a_0(x) = 1$, $\sigma_0(x) = 2\sqrt{|x|}$, and $y_0 = x_0^2$, the stochastic process $\zeta_T(t) = \xi_T^2(t)$ converges weakly, as $T \to +\infty$, to the solution ζ of the following Itô's equation:

$$\zeta(t) = x_0^2 + t + 2 \int_0^t \sqrt{\zeta(s)} \, d\widehat{W}(s). \tag{5.39}$$

Here $\zeta(t) \geq 0$ with probability 1 for all $t \geq 0$. Moreover, the assumptions of Theorem 5.8 with

$$\widehat{g}_T(x) = \left(\frac{b_T}{1 + b_T^2 |x|}\right)^{\frac{1}{2}}, \quad g_0(x) = \frac{\pi}{4\sqrt{|x|}}$$

hold for the functions $g_T(x)$, defined in Example 5.7.

Thus, by Theorem 5.8, the stochastic processes $\beta_T^{(2)}(t)$, defined by relation (5.38) converge weakly, as $T \to +\infty$, to the process $W^*\left(\beta^{(1)}\right)$, where

$$\beta^{(1)}(t) = \pi \left[\sqrt{\zeta(t)} - |x_0| - \widehat{W}(t)\right],$$

$\left(\zeta, \widehat{W}\right)$ is a solution of Eq. (5.39), and W^* is a Wiener process, W^* and $\beta^{(1)}$ are independent.

5.8   Auxiliary Results

Lemma 5.1 *Let ξ_T be a solution of Eq. (5.1) belonging to the class $K(G_T)$. Then for any $N > 0$ and for any Borel set $B \subset [-N; N]$ there exists a constant C_L such that*

$$\int_0^L P\{G_T(\xi_T(s)) \in B\}\, ds \leq C_L \psi(\lambda(B)),$$

where $\lambda(B)$ is the Lebesgue measure of the set B, $\psi(|x|)$ is a certain bounded function satisfying $\psi(|x|) \to 0$ as $|x| \to 0$.

Proof Consider the function

$$\Phi_T(x) = 2 \int_0^x f_T'(u) \left(\int_0^u \frac{\chi_B(G_T(v))}{f_T'(v)}\, dv\right) du.$$

The function $\Phi_T(x)$ is continuous, the derivative $\Phi_T'(x)$ of this function is continuous and the second derivative $\Phi_T''(x)$ exists a.e. with respect to the Lebesgue measure and is locally bounded. Therefore, we can apply the Itô formula to the process $\Phi_T(\xi_T(t))$, where $\xi_T(t)$ is a solution of Eq. (5.1).

Furthermore, the equality

$$\Phi_T'(x)a_T(x) + \frac{1}{2}\Phi_T''(x) = \chi_B(G_T(x)),$$

holds a.e. with respect to the Lebesgue measure. Using the latter equality we conclude that

$$\int\limits_0^t \chi_B \left(\zeta_T(s)\right) ds = \Phi_T(\xi_T(t)) - \Phi_T(x_0) - \int\limits_0^t \Phi_T'(\xi_T(s)) \, dW_T(s), \qquad (5.40)$$

with probability 1 for all $t \geq 0$, where $\zeta_T(t) = G_T(\xi_T(t))$. Hence, using the properties of stochastic integrals, we obtain that

$$\int\limits_0^t \mathsf{P}\{\zeta_T(s) \in B\} \, ds = \mathsf{E} \left[\Phi_T(\xi_T(t)) - \Phi_T(x_0)\right]. \qquad (5.41)$$

According to condition (A_4) we have

$$|\Phi_T(x) - \Phi_T(x_0)| \leq C\psi \left(\lambda(B)\right) \left[1 + |G_T(x)|^m\right].$$

Hence, using inequality (5.4), we obtain that

$$|\mathsf{E} \left[\Phi_T(\xi_T(L)) - \Phi_T(x_0)\right]| \leq C_L \psi \left(\lambda(B)\right)$$

for a certain constant C_L. The latter inequality and equality (5.41) prove Lemma 5.1.
$\square$

Lemma 5.2 *Let ξ_T be a solution of Eq. (5.1) belonging to the class $K(G_T)$. If for measurable locally bounded functions $q_T(x)$ condition (A_5) holds true, then for any $L > 0$*

$$\sup_{0 \leq t \leq L} \left| \int\limits_0^t q_T(\xi_T(s)) \, ds \right| \overset{\mathsf{P}}{\to} 0,$$

as $T \to +\infty$.

Proof Consider the function

$$\Phi_T(x) = 2 \int\limits_0^x f_T'(u) \left(\int\limits_0^u \frac{q_T(v)}{f_T'(v)} \, dv \right) du.$$

The same arguments as used to obtain equality (5.40) yield that

$$\int\limits_0^t q_T(\xi_T(s)) \, ds = \Phi_T(\xi_T(t)) - \Phi_T(x_0) - \int\limits_0^t \Phi_T'(\xi_T(s)) \, dW_T(s). \qquad (5.42)$$

It is clear that for any constants $\varepsilon > 0$, $N > 0$, $L > 0$ the following inequalities hold true:

$$\mathsf{P}\left\{\sup_{0 \le t \le L} |\Phi_T(\xi_T(t)) - \Phi_T(x_0)| > \varepsilon\right\} \le P_{NT} + \frac{4}{\varepsilon}\int\limits_{-N}^{N} f_T'(u)\left|\int\limits_0^u \frac{q_T(v)}{f_T'(v)}\,dv\right|\,du,$$

$$\mathsf{P}\left\{\sup_{0 \le t \le L}\left|\int\limits_0^t \Phi_T'(\xi_T(s))\,dW_T(s)\right| > \varepsilon\right\}$$

$$\le P_{NT} + \frac{4}{\varepsilon^2}\mathsf{E}\sup_{0 \le t \le L}\left|\int\limits_0^t \Phi_T'(\xi_T(s))\chi_{|\xi_T(s)|\le N}\,dW_T(s)\right|^2$$

$$\le P_{NT} + \frac{16}{\varepsilon^2}\mathsf{E}\int\limits_0^L \left[\Phi_T'(\xi_T(s))\right]^2 \chi_{|\xi_T(s)|\le N}\,ds,$$

where $P_{NT} = \mathsf{P}\left\{\sup\limits_{0 \le t \le L} |\xi_T(t)| > N\right\}$.

Using equality (5.42) and the properties of stochastic integrals, we conclude that

$$\mathsf{E}\int\limits_0^L \left[\Phi_T'(\xi_T(s))\right]^2 \chi_{|\xi_T(s)|\le N}\,ds = \mathsf{E}\left[\tilde{\Phi}_T(\xi_T(L)) - \tilde{\Phi}_T(x_0)\right], \tag{5.43}$$

where

$$\tilde{\Phi}_T(x) - \tilde{\Phi}_T(x_0) = 2\int\limits_{x_0}^x f_T'(u)\left(\int\limits_0^u \frac{1}{f_T'(v)}\left[\Phi_T'(v)\right]^2 \chi_{|v|\le N}\,dv\right)\,du.$$

Using the Hölder inequality, we obtain the following bounds:

$$|\tilde{\Phi}_T(x) - \tilde{\Phi}_T(x_0)| \le 2\left|\int\limits_{x_0}^x f_T'(u)\left|\int\limits_0^u \frac{1}{|f_T'(v)|^q}\,dv\right|^{\frac{1}{q}}\left|\int\limits_0^u \left[\Phi_T'(v)\right]^{2p}\chi_{|v|\le N}\,dv\right|^{\frac{1}{p}}\,du\right|$$

$$\le 2\left|\int\limits_{-N}^N |\Phi_T'(v)|^{2p}\,dv\right|^{\frac{1}{p}}\left|\int\limits_{x_0}^x f_T'(u)\left|\int\limits_0^u \frac{1}{|f_T'(v)|^q}\,dv\right|^{\frac{1}{q}}\,du\right|,$$

where $p > 1$ and $q > 1$ are arbitrary constant with $\frac{1}{p} + \frac{1}{q} = 1$.

Taking into account assumption (A_3) for certain $q = 1 + \delta$, we conclude that

$$
\left| \mathsf{E}\left[\tilde{\Phi}_T (\xi_T(L)) - \tilde{\Phi}_T (x_0) \right] \right| \leq 2C_L \left(\int\limits_{-N}^{N} |\Phi_T'(v)|^{2p} \, dv \right)^{\frac{1}{p}},
\tag{5.44}
$$

where the constants $C_L > 0$ and $p > 1$ do not depend on T.

According to assumption (A_5) and to Lebesgue's dominated convergence theorem, we have for arbitrary $N > 0$ that

$$
\int\limits_{-N}^{N} |\Phi_T'(v)|^{2p} \, dv \to 0,
$$

as $T \to +\infty$. Hence, taking into account the inequality (5.44) and the equality (5.43), we obtain the convergence

$$
\mathsf{E} \int\limits_{0}^{L} \left[\Phi_T' (\xi_T(s)) \right]^2 \chi_{|\xi_T(s)| \leq N} \, ds \to 0,
$$

as $T \to +\infty$.

According to assumptions (A_1) and (A_5), using Lebesgue's dominated convergence theorem, we obtain that

$$
\sup_{0 \leq t \leq L} |\Phi_T (\xi_T(t)) - \Phi_T(x_0)| \xrightarrow{\mathsf{P}} 0,
$$

$$
\sup_{0 \leq t \leq L} \left| \int\limits_{0}^{t} \Phi_T'(\xi_T(s)) \, dW_T(s) \right| \xrightarrow{\mathsf{P}} 0,
$$

as $T \to +\infty$. Thus, the equality (5.42) implies the statement of Lemma 5.2. $\square$

Remark 5.2 Let the assumptions of Lemma 5.2 hold. Suppose additionally that the measurable locally bounded functions g_T satisfy assumption (A_7). Then for any $L > 0$

$$
\mathsf{E} \sup_{0 \leq t \leq L} \left| \int\limits_{0}^{t} q_T (\xi_T(s)) \, ds \right| \to 0,
$$

as $T \to +\infty$.

Proof Using equality (5.42) we have that

$$
\sup_{0 \le t \le L} \left| \int_0^t q_T(\xi_T(s)) \, ds \right|^2 \le 2 \sup_{0 \le t \le L} |\Phi_T(\xi_T(t)) - \Phi_T(x_0)|^2
$$

$$
+ 2 \sup_{0 \le t \le L} \left| \int_0^t \Phi_T'(\xi_T(s)) \, dW_T(s) \right|^2 .
$$

Taking into account the first inequality from (A_7) we conclude

$$
|\Phi_T(x) - \Phi_T(x_0)|^2 = 2 \left| \int_{x_0}^x f_T'(u) \left(\int_0^u \frac{q_T(v)}{f_T'(v)} \, dv \right) du \right|^2
$$

$$
\le 2 \left| \int_{x_0}^x C \left[1 + |u|^{\alpha_1} \right] du \right|^2 \le \widehat{C} \left[1 + |x|^{\alpha_1 + 1} \right]^2 .
$$

Therefore,

$$
\mathsf{E} \sup_{0 \le t \le L} |\Phi_T(\xi_T(t)) - \Phi_T(x_0)|^2 \le C_L .
$$

Using condition (A_7) and the properties of stochastic integrals, we conclude that

$$
\mathsf{E} \sup_{0 \le t \le L} \left| \int_0^t \Phi_T'(\xi_T(s)) \, dW_T(s) \right|^2 \le 4 \mathsf{E} \int_0^L \left[\Phi_T'(\xi_T(s)) \right]^2 ds \le \widetilde{C}_L .
$$

Therefore,

$$
\mathsf{E} \sup_{0 \le t \le L} \left| \int_0^t q_T(\xi_T(s)) \, ds \right|^2 \le \widetilde{\widetilde{C}}_L .
$$

Consequently, for any $L > 0$

$$
\mathsf{E} \sup_{0 \le t \le L} \left| \int_0^t q_T(\xi_T(s)) \, ds \right| \to 0,
$$

as $T \to +\infty$. $\square$

Lemma 5.3 *Let ξ_T be a solution of Eq. (5.1) belonging to the class $K(G_T)$. Also let $\zeta_T(t) = G_T(\xi_T(t)) \xrightarrow{P} \zeta(t)$, as $T \to +\infty$. Then for any measurable locally bounded function g the following convergence holds true:*

$$\sup_{0 \le t \le L} \left| \int_0^t g(\zeta_T(s))\, ds - \int_0^t g(\zeta(s))\, ds \right| \xrightarrow{P} 0,$$

as $T \to +\infty$, for any constant $L > 0$.

Proof Let $\varphi_N(x) = 1$ for $|x| \le N$, $\varphi_N(x) = N + 1 - |x|$ for $|x| \in [N, N+1]$ and $\varphi_N(x) = 0$ for $|x| > N + 1$. Then for all $T > 0, L > 0$

$$P\left\{ \sup_{0 \le t \le L} \left| \int_0^t [g(\zeta_T(s)) - g(\zeta_T(s))\varphi_N(\zeta_T(s))]\, ds \right| > 0 \right\} \le P\left\{ \sup_{0 \le t \le L} |\zeta_T(t)| > N \right\},$$

$$P\left\{ \sup_{0 \le t \le L} \left| \int_0^t [g(\zeta(s)) - g(\zeta(s))\varphi_N(\zeta(s))]\, ds \right| > 0 \right\} \le P\left\{ \sup_{0 \le t \le L} |\zeta(t)| > N \right\}$$

$$\le \limsup_{T \to +\infty} P\left\{ \sup_{0 \le t \le L} |\zeta_T(t)| > N \right\}.$$

According to Theorem 5.7 the convergence (5.10) holds for the process ζ_T. So, to complete the proof of Lemma 5.3, we need to establish that

$$\int_0^L |g(\zeta_T(s))\varphi_N(\zeta_T(s)) - g(\zeta(s))\varphi_N(\zeta(s))|\, ds \xrightarrow{P} 0 \qquad (5.45)$$

as $T \to +\infty$.

First, assume that the function g is continuous. Then

$$g(\zeta_T(s))\varphi_N(\zeta_T(s)) - g(\zeta(s))\varphi_N(\zeta(s)) \xrightarrow{P} 0$$

as $T \to +\infty$ for all $0 \le s \le L$ and $|g(x)\varphi_N(x)| \le C_N$ for all x. Thus, according to Lebesgue's dominated convergence theorem, we have the convergence (5.45). Second, let the function g be measurable and locally bounded. Then, using Luzin's theorem we conclude that for any $\delta > 0$ there exists a continuous function $g^\delta(x)$, which coincides with $g(x)$ for $x \notin B^\delta$, where $B^\delta \subset [-N-1, N+1]$ and its

Lebesgue measure satisfies inequality $\lambda\left(B^\delta\right) < \delta$. Thus, for every $\delta > 0$ the relation (5.45) holds true for the function $g^\delta(x)$. Since for any $\varepsilon > 0$

$$P\left\{\int_0^L \left|g(\zeta_T(s))\varphi_N(\zeta_T(s)) - g^\delta(\zeta_T(s))\varphi_N(\zeta_T(s))\right| ds > \varepsilon\right\}$$

$$\leq \frac{2}{\varepsilon}\mathsf{E}\int_0^L \left|g(\zeta_T(s))\varphi_N(\zeta_T(s)) - g^\delta(\zeta_T(s))\varphi_N(\zeta_T(s))\right| \chi_{\{B^\delta\}}(\zeta_T(s))\,ds$$

$$\leq \frac{C_N}{\varepsilon}\int_0^L \mathsf{P}\left\{\zeta_T(s) \in B^\delta\right\} ds,$$

and

$$P\left\{\int_0^L \left|g(\zeta(s))\varphi_N(\zeta(s)) - g^\delta(\zeta(s))\varphi_N(\zeta(s))\right| ds > \varepsilon\right\}$$

$$\leq \frac{C_N}{\varepsilon}\int_0^L \mathsf{P}\left\{\zeta(s) \in B^\delta\right\} ds \leq \frac{C_N}{\varepsilon}\limsup_{T\to+\infty}\int_0^L \mathsf{P}\left\{\zeta_T(s) \in B^\delta\right\} ds,$$

we can additionally take into account Lemma 5.1, and conclude that the relation (5.45) holds for a measurable and locally bounded function g, as well.   □

Lemma 5.4 *Let ξ_T be a solution of Eq. (5.1) belonging to the class $K\left(G_T\right)$ and let $\zeta_T(t) = G_T(\xi_T(t)) \overset{P}{\to} \zeta(t)$, $\eta_T(t) = \int_0^t G_T'(\xi_T(s))\,dW_T(s) \overset{P}{\to} \eta(t)$, as $T \to +\infty$. Then for a measurable locally bounded function g the following convergence holds true:*

$$\sup_{0\leq t\leq L}\left|\int_0^t g(\zeta_T(s))\,d\eta_T(s) - \int_0^t g(\zeta(s))\,d\eta(s)\right| \overset{P}{\to} 0$$

as $T \to +\infty$ for any constant $L > 0$.

Proof Similarly to the proof of Lemma 5.3, it is sufficient to obtain an analog of the convergence (5.45), i.e., to get that for any $N > 0, L > 0$

$$\sup_{0 \leq t \leq L} \left| \int_0^t g(\zeta_T(s)) \varphi_N(\zeta_T(s)) \, d\eta_T(s) - \int_0^t g(\zeta(s)) \varphi_N(\zeta(s)) \, d\eta(s) \right| \xrightarrow{P} 0$$

$$(5.46)$$

as $T \to +\infty$, where $\varphi_N(x)$ is defined in the proof of Lemma 5.3. The proof of the convergence (5.46) for a continuous function $g(x)$ is similar to the proof of the corresponding theorem in [79, Chap.2, §6]. The explicit form of the quadratic characteristic $\langle \eta_T \rangle(t)$ of the martingale $\eta_T(t)$ and condition (A_1) imply the inequality

$$\int_0^L [\varphi_N(\zeta_T(t))]^2 \, d\langle \eta_T \rangle(t) \leq C_N L,$$

which is used for the proof of the convergence (5.46). The extension of such a convergence to the class of measurable locally bounded functions is based on Lemma 5.1 and is provided similarly to the proof of Lemma 5.3. □

Lemma 5.5 *Let ξ_T be a solution of Eq. (5.1) belonging to the class K (G_T), and let the stochastic process $(\zeta_T(t), \eta_T(t))$, with $\zeta_T(t) = G_T(\xi_T(t))$ and $\eta_T(t) = \int_0^t G_T'(\xi_T(s)) \, dW_T(s)$ be stochastically equivalent to the process $(\widetilde{\zeta}_T(t), \widetilde{\eta}_T(t))$. Then the process*

$$\int_0^t g(\zeta_T(s)) \, ds + \int_0^t q(\zeta_T(s)) \, d\eta_T(s),$$

where g, q are measurable locally bounded functions, is stochastically equivalent to the process

$$\int_0^t g(\widetilde{\zeta}_T(s)) \, ds + \int_0^t q(\widetilde{\zeta}_T(s)) \, d\widetilde{\eta}_T(s).$$

Proof The proof is the same as that of Corollary A.3 from the Appendix. □

Chapter 6
Asymptotic Behavior of Homogeneous Additive Functionals of the Solutions to Inhomogeneous Itô SDEs with Non-regular Dependence on a Parameter

In this chapter, we consider the asymptotic behavior, as $T \to +\infty$, of some functionals of the form $I_T(t) = F_T(\xi_T(t)) + \int_0^t g_T(\xi_T(s)) \, dW_T(s)$, $t \geq 0$. Here $\xi_T(t)$ is the solution to the time-inhomogeneous Itô stochastic differential equation

$$d\xi_T(t) = a_T(t, \xi_T(t)) \, dt + dW_T(t), \quad t \geq 0, \quad \xi_T(0) = x_0,$$

where $T > 0$ is a parameter, $a_T(t, x)$, $x \in \mathbb{R}$ are measurable functions, $|a_T(t, x)| \leq L_T$ for all $x \in \mathbb{R}$ and $t \geq 0$, W_T are standard Wiener processes, $F_T(x)$, $x \in \mathbb{R}$ are continuous functions, and $g_T(x)$, $x \in \mathbb{R}$ are measurable locally bounded functions. Section 6.1 contains some preliminary remarks, notations and basic definitions. The asymptotic behavior of the integral functionals of the Lebesgue integral type is investigated in Sect. 6.3. Section 6.4 contains some results about the weak convergence of the martingale type functionals and the mixed functionals. Section 6.5 includes several examples. Auxiliary results are collected in Sect. 6.6.

6.1 Preliminaries

Consider the time-inhomogeneous Itô stochastic differential equation

$$d\xi_T(t) = a_T(t, \xi_T(t)) \, dt + dW_T(t), \quad t \geq 0, \ \xi_T(0) = x_0, \tag{6.1}$$

where $T > 0$ is a parameter, $a_T(t, x)$, $x \in \mathbb{R}$ are real-valued measurable functions such that $|a_T(t, x)| \leq L_T$ for all (t, x) and some family of constants $L_T > 0$, and $W_T = \{W_T(t), t \geq 0\}$, $T > 0$ is a family of standard Wiener processes defined on a complete probability space $(\Omega, \mathfrak{I}, \mathsf{P})$.

© Springer Nature Switzerland AG 2020
G. Kulinich et al., *Asymptotic Analysis of Unstable Solutions of Stochastic Differential Equations*, Bocconi & Springer Series 9,
https://doi.org/10.1007/978-3-030-41291-3_6

It is known from Theorem 4 in [82] that for any $T > 0$ and $x_0 \in \mathbb{R}$ Eq. (5.1) possesses a unique strong solution $\xi_T = \{\xi_T(t), t \geq 0\}$.

In this chapter, we study the weak convergence, as $T \to +\infty$, of the processes $I_T(t) = F_T(\xi_T(t)) + \int_0^t g_T(\xi_T(s)) \, dW_T(s)$, where $\xi_T(t)$ is the solution to the stochastic differential equation (6.1), $F_T(x)$ is a family of continuous real-valued functions, and $g_T(x)$ is a family of measurable locally bounded real-valued functions. All the results about asymptotic behavior are obtained under the condition which provides a certain proximity of the coefficients $a_T(t, x)$ to some measurable functions $\widehat{a}_T(x)$. In a such situation, the limit processes, obtained when $T \to +\infty$, are some functionals of the limits of the solutions $\widehat{\xi}_T(t)$ to the homogeneous stochastic differential equations

$$d\widehat{\xi}_T(t) = \widehat{a}_T(\widehat{\xi}_T(t)) \, dt + dW_T(t). \tag{6.2}$$

The present chapter generalizes similar results from the previous chapter, that were formulated for the unique strong solutions $\widehat{\xi}_T$ to the homogeneous stochastic differential equations (6.2), to the case of the solutions $\xi_T(t)$ to inhomogeneous equations (6.1). Under the proposed conditions, we prove that the asymptotic behavior of the solutions and some functionals of the solutions to the inhomogeneous Itô stochastic differential equations (6.1) is the same as that for the solutions to the homogeneous Itô stochastic differential equations (6.2). We assume that the drift coefficient $a_T(t, x)$ in Eq. (5.1) can have non-regular dependence on the parameter. For example, the drift coefficient $a_T(t, x)$ can tend, as $T \to +\infty$, to infinity at some points x_k and at some points t_k as well, or it can have degeneracies of some other types.

In what follows we denote by C, L, N, C_N, L_N any constants that do not depend on T, x nor t. To formulate and prove the main results, we introduce functions of the form

$$f_T(x) = \int_0^x \exp\left\{-2 \int_0^u \widehat{a}_T(v) \, dv\right\} du, \quad T > 0. \tag{6.3}$$

Throughout the chapter we use the following notations:

$$\beta_T^{(1)}(t) = \int_0^t g_T(\xi_T(s)) \, ds, \quad \beta_T^{(2)}(t) = \int_0^t g_T(\xi_T(s)) \, dW_T(s),$$

$$I_T(t) = F_T(\xi_T(t)) + \int_0^t g_T(\xi_T(s)) \, dW_T(s),$$

where ξ_T and W_T are related via Eq. (6.1), g_T is a family of measurable, locally bounded real-valued functions, and F_T is a family of continuous real-valued functions.

To study the weak convergence, as $T \to +\infty$, of the processes $I_T(t) = F_T(\xi_T(t)) + \int_0^t g_T(\xi_T(s))\,dW_T(s)$, where ξ_T is the solution to the stochastic differential equation (6.1), we suppose additionally that the drift coefficients satisfy the following assumption:

There exists a family of measurable, locally bounded functions $\widehat{a}_T(x)$ such that for any $L > 0$

$$(A_0) \qquad\qquad \lim_{T \to +\infty} \int_0^L \sup_x |a_T(t, x) - \widehat{a}_T(x)|\,dt = 0.$$

Note that, due to condition (A_0), some results about solutions $\widehat{\xi}_T$ to the homogeneous equations (6.2), which are obtained in Chap. 5, can be extended to solutions ξ_T to the inhomogeneous equations (6.1). Therefore, by analogy to Chap. 5, we consider Eq. (6.1) belonging to the class $K(G_T)$.

Definition 6.1 The class of equations of the form (6.1) will be denoted by $K(G_T)$, if there exist families of functions $\widehat{a}_T(x)$ and $G_T(x)$, $x \in \mathbb{R}$, such that:

(1) $\widehat{a}_T(x)$ are measurable locally bounded real-valued functions, satisfying condition (A_0);
(2) $G_T(x)$ have continuous derivatives $G_T'(x)$ and locally integrable second derivatives $G_T''(x)$ a.e. with respect to the Lebesgue measure such that, for all $T > 0$, $x \in \mathbb{R}$ and $t \geq 0$, for some constant $C > 0$ the following inequalities hold:

$$(A_1) \qquad \left[G_T'(x)a_T(t, x) + \tfrac{1}{2}G_T''(x)\right]^2 + \left[G_T'(x)\right]^2 \leq C\left[1 + |G_T(x)|^2\right],$$
$$|G_T(x_0)| \leq C;$$

(3) there exist constants $C > 0$ and $\alpha > 0$ such that, for all $x \in \mathbb{R}$,

$$|G_T(x)| \geq C|x|^\alpha;$$

(4) there exist a bounded function $\psi(|x|)$ and a constant $m \geq 0$ such that $\psi(|x|) \to 0$, as $|x| \to 0$, and, for all $x \in \mathbb{R}$ and $T > 0$ and for any measurable bounded set B, the following inequality holds:

$$(A_2) \qquad \left| f_T'(x) \int_0^x \frac{\chi_B(G_T(u))}{f_T'(u)}\,du \right| \leq \psi(\lambda(B))\left[1 + |x|^m\right],$$

where $\chi_B(x)$ is the indicator function of a set B, $\lambda(B)$ is the Lebesgue measure of B, and $f_T'(x)$ is the derivative of the function $f_T(x)$ defined by equality (6.3).

Assume that, for certain locally bounded functions $q_T(x)$ and any constant $N > 0$, the following condition holds:

$$(A_3) \qquad \lim_{T \to +\infty} \sup_{|x| \le N} \left| f_T'(x) \int_0^x \frac{q_T(v)}{f_T'(v)} \, dv \right| = 0.$$

6.2   Weak Compactness and Weak Convergence of the Solutions of Itô SDEs

Now we are in a position to use the result concerning the weak compactness of stochastic processes $\zeta_T = \{\zeta_T(t) = G_T(\xi_T(t)), t \ge 0\}$ (Theorem 6.1) in further investigation of asymptotic behavior of the solutions (Theorem 6.2) and some functionals of the solutions (Theorems 6.3–6.7) to the inhomogeneous Itô stochastic differential equations (6.1). In the proofs of the theorems, which are performed similarly to the proofs of the corresponding theorems in Chap. 5, we emphasize the differences associated with the inhomogeneous equations.

Theorem 6.1 *Let ξ_T be a solution of Eq. (6.1) and let there exist a family of continuous functions $G_T(x)$, $x \in \mathbb{R}$ with continuous derivative $G_T'(x)$ and the second derivative $G_T''(x)$ to be assumed to exist a.e. with respect to the Lebesgue measure and to be locally integrable. Let the functions $G_T(x)$ satisfy assumption (A_1), for all $T > 0$, $t \ge 0$, $x \in \mathbb{R}$. Then the family of the processes $\zeta_T = \{\zeta_T(t) = G_T(\xi_T(t)), t \ge 0\}$ is weakly compact.*

Proof The proof of Theorem 6.1 differs from the proof of Theorem 5.1 only by the representation of the process $\zeta_T(t) = G_T(\xi_T(t))$. Now we have that

$$\zeta_T(t) = G_T(x_0) + \int_0^t L_T(\xi_T(s)) \, ds + \int_0^t G_T'(\xi_T(s)) \, dW_T(s), \qquad (6.4)$$

where

$$L_T(x) = G_T'(x) a_T(t, x) + \frac{1}{2} G_T''(x),$$

comparing to formula (5.3).   □

Theorem 6.2 *Let ξ_T be a solution of Eq. (6.1) belonging to the class $K(G_T)$ and $G_T(x_0) \to y_0$, as $T \to +\infty$. Assume that there exist measurable locally bounded functions $a_0(x)$ and $\sigma_0(x)$ such that:*

(1) *the functions*

$$q_T^{(1)}(x) = G_T'(x)\widehat{a}_T(x) + \frac{1}{2} G_T''(x) - a_0\left(G_T(x)\right),$$

$$q_T^{(2)}(x) = \left[G_T'(x)\right]^2 - \sigma_0^2\left(G_T(x)\right),$$

satisfy assumption (A_3);

(2) *the Itô equation (5.14) has a unique weak solution* $\left(\zeta,\, \widehat{W}\right)$.

Then the stochastic processes $\zeta_T = G_T(\xi_T(t))$ *converge weakly, as* $T \to +\infty$, *to the solution* ζ *of Eq. (5.14).*

Proof Rewrite the equality (6.4) as

$$\zeta_T(t) = G_T(x_0) + \int_0^t a_0(\zeta_T(s))\, ds + \eta_T(t) + \alpha_T^{(0)}(t) + \alpha_T^{(1)}(t), \qquad (6.5)$$

where

$$\eta_T(t) = \int_0^t G_T'(\xi_T(s))\, dW_T(s),$$

$$\alpha_T^{(0)}(t) = \int_0^t G_T'(\xi_T(s))\Delta a_T(s)\, ds, \quad \Delta a_T(s) = a_T(s, \xi_T(s)) - \widehat{a}_T(\xi_T(s)),$$

$$\alpha_T^{(1)}(t) = \int_0^t q_T^{(1)}(\xi_T(s))\, ds, \quad q_T^{(1)}(x) = G_T'(x)\widehat{a}_T(x) + \frac{1}{2} G_T''(x) - a_0\left(G_T(x)\right).$$

The conditions (A_0) and (A_1), together with the inequality (5.9), imply that

$$\sup_{0 \le t \le L} \left|\alpha_T^{(0)}(t)\right| \le \int_0^L \left|G_T'(\xi_T(s))\right| \, |\Delta a_T(s)|\, ds$$

$$\le \left[C\left(1 + \sup_{0 \le s \le L} |\zeta_T(s)|^2\right)\right]^{\frac{1}{2}} \int_0^L \sup_x |a_T(s, x) - \widehat{a}_T(x)|\, ds \overset{P}{\to} 0, \qquad (6.6)$$

as $T \to +\infty$, for any $L > 0$.

The functions $q_T^{(1)}(x)$ satisfy the conditions of Lemma 6.2. Thus, for any $L > 0$

$$\sup_{0 \le t \le L} \left| \alpha_T^{(1)}(t) \right| \overset{\mathsf{P}}{\to} 0 \qquad (6.7)$$

as $T \to +\infty$.

It is clear that $\eta_T(t)$ is a family of continuous martingales with the quadratic characteristics

$$\langle \eta_T \rangle(t) = \int_0^t \left[G_T'(\xi_T(s)) \right]^2 ds = \int_0^t \sigma_0^2(\zeta_T(s)) \, ds + \alpha_T^{(2)}(t), \qquad (6.8)$$

where

$$\alpha_T^{(2)}(t) = \int_0^t q_T^{(2)}(\xi_T(s)) \, ds, \quad q_T^{(2)}(x) = \left(G_T'(x) \right)^2 - \sigma_0^2 \left(G_T(x) \right).$$

The functions $q_T^{(2)}(x)$ satisfy the conditions of Lemma 6.2. Thus, for any $L > 0$

$$\sup_{0 \le t \le L} \left| \alpha_T^{(2)}(t) \right| \overset{\mathsf{P}}{\to} 0, \qquad (6.9)$$

as $T \to +\infty$.

According to Theorem 6.1, the family of the processes $\zeta_T(t)$ is weakly compact. It is easy to see that the compactness conditions (5.12) are fulfilled for the processes $\eta_T(t)$. Using the convergences (6.6), (6.7), and (6.9), we have that relations (5.12) hold for the processes $\alpha_T^{(k)}(t)$, $k = 0, 1, 2$, as well. It means that we can apply Skorokhod's convergent subsequence principle (see Theorem A.12) for the process

$$\left(\zeta_T(t), \eta_T(t), \alpha_T^{(k)}(t), \ k = 0, 1, 2 \right).$$

According to this principle, given an arbitrary sequence $T_n' \to +\infty$, we can choose a subsequence $T_n \to +\infty$, a probability space $(\tilde{\Omega}, \tilde{\mathfrak{S}}, \tilde{\mathsf{P}})$, and a stochastic process

$$\left(\tilde{\zeta}_{T_n}(t), \tilde{\eta}_{T_n}(t), \tilde{\alpha}_{T_n}^{(k)}(t), \ k = 0, 1, 2 \right)$$

defined on this space such that its finite-dimensional distributions coincide with those of the process

$$\left(\zeta_{T_n}(t), \eta_{T_n}(t), \alpha_{T_n}^{(k)}(t), \ k = 0, 1, 2 \right)$$

and, moreover,

$$\tilde{\zeta}_{T_n}(t) \overset{\tilde{P}}{\to} \tilde{\zeta}(t), \quad \tilde{\eta}_{T_n}(t) \overset{\tilde{P}}{\to} \tilde{\eta}(t), \quad \tilde{\alpha}^{(k)}_{T_n}(t) \overset{\tilde{P}}{\to} \tilde{\alpha}^{(k)}(t), \ k = 0, 1, 2,$$

as $T_n \to +\infty$, for all $0 \le t \le L$, where $\tilde{\zeta}(t)$, $\tilde{\eta}(t)$, $\tilde{\alpha}^{(k)}(t)$, $k = 0, 1, 2$ are some stochastic processes.

Evidently, relations (6.6)–(6.9) imply that $\tilde{\alpha}^{(k)}(t) \equiv 0, k = 0, 1, 2$ a.s. According to (5.10), the processes $\tilde{\zeta}(t)$ and $\tilde{\eta}(t)$ are continuous with probability 1. Moreover, applying Lemma 5.5 together with equalities (6.5) and (6.8), we obtain that

$$\tilde{\zeta}_{T_n}(t) = G_{T_n}(x_0) + \int_0^t a_0(\tilde{\zeta}_{T_n}(s))\, ds + \tilde{\alpha}^{(0)}_{T_n}(t) + \tilde{\alpha}^{(1)}_{T_n}(t) + \tilde{\eta}_{T_n}(t), \tag{6.10}$$

$$\langle \tilde{\eta}_{T_n}\rangle(t) = \int_0^t \sigma_0^2(\tilde{\zeta}_{T_n}(s))\, ds + \tilde{\alpha}^{(2)}_{T_n}(t),$$

where

$$\tilde{\zeta}_{T_n}(t) \overset{\tilde{P}}{\to} \tilde{\zeta}(t), \quad \tilde{\eta}_{T_n}(t) \overset{\tilde{P}}{\to} \tilde{\eta}(t), \quad \sup_{0 \le t \le L} \left| \tilde{\alpha}^{(k)}_{T_n}(t) \right| \overset{\tilde{P}}{\to} 0, \quad k = 0, 1, 2$$

as $T_n \to +\infty$.

An analog of the convergence (5.12) holds for the processes $\tilde{\zeta}_{T_n}(t)$ and $\tilde{\eta}_{T_n}(t)$. Therefore, according to the well-known result of Prokhorov (see Theorem A.13), we conclude that for any $L > 0$

$$\sup_{0 \le t \le L} \left| \tilde{\zeta}_{T_n}(t) - \tilde{\zeta}(t) \right| \overset{\tilde{P}}{\to} 0, \quad \sup_{0 \le t \le L} \left| \tilde{\eta}_{T_n}(t) - \tilde{\eta}(t) \right| \overset{\tilde{P}}{\to} 0 \tag{6.11}$$

as $T_n \to +\infty$.

According to Lemma 5.3, we can pass to the limit in (6.10) and obtain the representation

$$\tilde{\zeta}(t) = y_0 + \int_0^t a_0(\tilde{\zeta}(s))\, ds + \tilde{\eta}(t),$$

where $\tilde{\eta}(t)$ is the almost surely continuous martingale with the quadratic characteristic

$$\langle \tilde{\eta}\rangle(t) = \int_0^t \sigma_0^2(\tilde{\zeta}(s))\, ds.$$

Now, it is well known that the latter representation provides the existence of a Wiener process $\widehat{W}$ such that

$$\tilde{\eta}(t) = \int_0^t \sigma_0(\tilde{\zeta}(s)) \, d\widehat{W}(s).$$

Thus, the process $\left(\tilde{\zeta}, \widehat{W}\right)$ satisfies Eq. (5.14), and the processes $\tilde{\zeta}_{T_n}(t)$ converge weakly, as $T_n \to +\infty$, to the process $\tilde{\zeta}$. Since the sequence $T_n' \to +\infty$ is arbitrary, and since the solution to Eq. (5.14) is weakly unique, the proof of Theorem 6.2 is complete. $\qquad\square$

6.3   Asymptotic Behavior of Integral Functionals of the Lebesgue Integral Type

In this section we obtain sufficient conditions for the weak convergence of some integral functionals of the Lebesgue integral type.

Theorem 6.3 *Let ξ_T be a solution of Eq. (6.1) belonging to the class $K(G_T)$ and let assumptions of Theorem 6.2 hold. Assume that for measurable and locally bounded functions g_T there exists a measurable and locally bounded function g_0 such that the function*

$$q_T(x) = g_T(x) - g_0(G_T(x))$$

satisfies assumption (A_3). Then the stochastic processes $\beta_T^{(1)}(t) = \int_0^t g_T(\xi_T(s)) \, ds$ converge weakly, as $T \to +\infty$, to the process

$$\beta^{(1)}(t) = \int_0^t g_0(\zeta(s)) \, ds,$$

where ζ is the solution of Eq. (5.14).

The proof of Theorem 6.3 is literally the same as that of Theorem 5.3.

Theorem 6.4 *Let ξ_T be a solution of Eq. (6.1) belonging to the class $K(G_T)$, and let the assumptions of Theorem 6.2 hold. Assume that, for measurable and locally bounded functions g_T, there exists a measurable locally bounded function g_0 such that*

$$\left| f_T'(x) \int_0^x \frac{g_T(v)}{f_T'(v)} \, dv \right| \chi_{|x| \le N} \le C_N,$$

(A_4)
$$\lim_{T \to +\infty} \sup_{|x| \le N} \left| f_T'(x) \int\limits_0^x \frac{g_T(v)}{f_T'(v)} \, dv - g_0 \left(G_T(x) \right) G_T'(x) \right| = 0$$

for all $N > 0$. Then the stochastic processes

$$\beta_T^{(1)}(t) = \int\limits_0^t g_T(\xi_T(s)) \, ds$$

converge weakly, as $T \to +\infty$, to the process

$$\tilde{\beta}^{(1)}(t) = 2 \left(\int\limits_{y_0}^{\zeta(t)} g_0(x) \, dx - \int\limits_0^t g_0(\zeta(s)) \, \sigma_0(\zeta(s)) \, d\widehat{W}(s) \right),$$

where $\left(\zeta, \, \widehat{W} \right)$ is the solution to Eq. (5.14).

Proof The proof of Theorem 6.4 differs from the proof of Theorem 5.3 only in that we use the different representation of the functional $\beta_T^{(1)}(t) = \int_0^t g_T(\xi_T(s)) \, ds$. In this case we have

$$\beta_T^{(1)}(t) = 2 \int\limits_{G_T(x_0)}^{\zeta_T(t)} g_0(u) \, du - 2 \int\limits_0^t g_0 \left(\zeta_T(s) \right) d\eta_T(s) + \gamma_T^{(1)}(t) - \gamma_T^{(2)}(t) - \gamma_T^{(0)}(t),$$

where

$$\gamma_T^{(1)}(t) = \int\limits_{x_0}^{\xi_T(t)} \widehat{q}_T(u) \, du, \qquad \gamma_T^{(2)}(t) = \int\limits_0^t \widehat{q}_T(\xi_T(s)) \, dW_T(s),$$

$$\gamma_T^{(0)}(t) = \int\limits_0^t \Phi_T'(\xi_T(s)) \left[a_T(s, \xi_T(s)) - \widehat{a}_T(\xi_T(s)) \right] ds,$$

$$\Phi_T(x) = 2 \int\limits_0^x f_T'(u) \left(\int\limits_0^u \frac{g_T(v)}{f_T'(v)} \, dv \right) du,$$

$$\widehat{q}_T(x) = \Phi_T'(x) - 2g_0 \left(G_T(x) \right) G_T'(x),$$

and $f_T'(x)$ is the derivative of the function $f_T(x)$ defined by the equality (6.3).

The latter representation differs from the corresponding representation in the proof of Theorem 5.3 by the term $\gamma_T^{(0)}(t)$. For any constants $\varepsilon > 0$, $N > 0$, and $L > 0$, we have the inequalities

$$
P\left\{ \sup_{0 \le t \le L} \left| \gamma_T^{(0)}(t) \right| > \varepsilon \right\} \le P_{NT}
$$

$$
+ \frac{2}{\varepsilon} \int_0^L E \left| \Phi_T'(\xi_T(s)) \right| \left| a_T(s, \xi_T(s)) - \widehat{a}_T(\xi_T(s)) \right| \chi_{|\xi_T(s)| \le N} \, ds
$$

$$
\le P_{NT} + \frac{2}{\varepsilon} C_N \int_0^L \sup_x \left[a_T(s, x) - \widehat{a}_T(x) \right] ds,
$$

where $P_{NT} = P\left\{ \sup_{0 \le t \le L} |\xi_T(t)| > N \right\}$. Using condition (3) from Definition 6.1 and the inequality (5.9), we obtain the convergence $\lim_{N \to +\infty} \limsup_{T \to +\infty} P_{NT} = 0$. Taking into account the assumptions of Theorem 6.4, we conclude that

$$
\sup_{0 \le t \le L} \left| \gamma_T^{(0)}(t) \right| \xrightarrow{P} 0 \tag{6.12}
$$

for any $L > 0$, as $T \to +\infty$. The rest of the proof of Theorem 6.4 is the same as that of Theorem 5.3. $\qquad\square$

Theorem 6.5 *Let ξ_T be a solution of Eq. (6.1) belonging to the class $K(G_T)$, and let the assumptions of Theorem 6.2 hold. Suppose that the functions $\widehat{a}_T$ satisfy assumption (A_3). Assume that, for measurable and locally bounded functions g_T, there exist two constants c_0 and b_0 such that for all $N > 0$*

$$
\left| f_T'(x) \int_0^x \frac{g_T(v)}{f_T'(v)} \, dv \right| \chi_{|x| \le N} \le C_N,
$$

$$
\lim_{T \to +\infty} \sup_{|x| \le N} \left| \int_0^x \left[f_T'(u) \int_0^u \frac{g_T(v)}{f_T'(v)} \, dv - c_0 \right] du \right| = 0,
$$

and the functions

$$
q_T(x) = \left[f_T'(x) \int_0^x \frac{g_T(v)}{f_T'(v)} \, dv - c_0 \right]^2 - b_0^2
$$

satisfy assumption (A_3).

Then the stochastic processes

$$\beta_T^{(1)}(t) = \int_0^t g_T\left(\xi_T(s)\right) ds$$

converge weakly, as $T \to +\infty$, *to the process* $2b_0 W(t)$, *where* W *is a Wiener process.*

Proof For the functional $\beta_T^{(1)}(t) = \int_0^t g_T(\xi_T(s)) ds$ we have the representation, for all $t \geq 0$, and with probability 1

$$\beta_T^{(1)}(t) = 2c_0 \int_0^t \widehat{a}_T(\xi_T(s)) ds + \gamma_T(t) - \eta_T^{(1)}(t) - \gamma_T^{(0)}(t) + \gamma_T^{(3)}(t),$$

where

$$\gamma_T(t) = 2 \int_{x_0}^{\xi_T(t)} \left[f_T'(u) \int_0^u \frac{g_T(v)}{f_T'(v)} dv - c_0 \right] du,$$

$$\eta_T^{(1)}(t) = \int_0^t \left[\Phi_T'(\xi_T(s)) - 2c_0 \right] dW_T(s),$$

and

$$\gamma_T^{(3)}(t) = 2c_0 \int_0^t \left[a_T(s, \xi_T(s)) - \widehat{a}_T(\xi_T(s)) \right] ds,$$

where $\gamma_T^{(0)}(t)$ and $\Phi_T(x)$ are defined in the proof of Theorem 6.4.

The functions $\widehat{a}_T$ satisfy condition (A_3). Thus, using Lemma 6.2, we conclude that for any $L > 0$

$$\sup_{0 \leq t \leq L} \left| \int_0^t \widehat{a}_T(\xi_T(s)) ds \right| \xrightarrow{\text{P}} 0,$$

as $T \to +\infty$.

For any constants $\varepsilon > 0$, $N > 0$ and $L > 0$, we have the inequalities

$$
P\left\{ \sup_{0 \le t \le L} |\gamma_T(t)| > \varepsilon \right\} \le P_{NT} + \frac{1}{\varepsilon} E \sup_{0 \le t \le L} \left| \int_{x_0}^{\xi_T(t)} \left[\Phi_T'(u) - 2c_0 \right] du \right| \chi_{|\xi_T(t)| \le N}
$$

$$
\le P_{NT} + \frac{2}{\varepsilon} N \sup_{|x| \le N} \left| \int_{x_0}^{x} \left[f_T'(u) \int_{0}^{u} \frac{g_T(v)}{f_T'(v)} dv - c_0 \right] du \right|,
$$

where P_{NT} is the same as that in the proof of Theorem 6.4. Using the latter inequality and the assumptions of Theorem 6.5, we conclude that

$$
\sup_{0 \le t \le L} |\gamma_T(t)| \xrightarrow{P} 0,
$$

as $T \to +\infty$.

Since the term $\gamma_T^{(0)}(t)$ is the same as that in the proof of Theorem 6.4, we get (6.12).

The inequality

$$
\sup_{0 \le t \le L} \left| \gamma_T^{(3)}(t) \right| \le 2|c_0| \int_{0}^{L} \sup_{x} \left[a_T(s, x) - \widehat{a}_T(x) \right] ds
$$

implies that for any $L > 0$

$$
\sup_{0 \le t \le L} \left| \gamma_T^{(3)}(t) \right| \xrightarrow{P} 0,
$$

as $T \to +\infty$.

Thus, we have that for any $L > 0$

$$
\sup_{0 \le t \le L} \left| \beta_T^{(1)}(t) + \eta_T^{(1)}(t) \right| \xrightarrow{P} 0,
$$

as $T \to +\infty$.

It is clear that $\eta_T^{(1)}(t)$ is the almost surely continuous martingale with the quadratic characteristic

$$
\langle \eta_T^{(1)} \rangle(t) = 4b_0^2 t + \int_{0}^{t} q_T(\xi_T(s)) ds,
$$

where $q_T(x) = \left[\Phi_T'(x) - 2c_0\right]^2 - 4b_0^2$. The functions $q_T(x)$ satisfy condition (A_3). Thus, using Lemma 6.2, we conclude that for any $L > 0$

$$\sup_{0 \le t \le L} \left|\langle \eta_T^{(1)} \rangle(t) - 4b_0^2 t\right| \overset{\mathsf{P}}{\to} 0,$$

as $T \to +\infty$.

Then, using the random time change in stochastic integrals (see, e.g., [74]), we obtain $\eta_T^{(1)}(t) = W_T^*\left(\langle \eta_T^{(1)} \rangle(t)\right)$, where $W_T^*(t)$ is a Wiener process. The same arguments as used to get (4.19) yield that

$$\sup_{0 \le t \le L} \left|\beta_T^{(1)}(t) - W_T^*\left(4b_0^2 t\right)\right| \overset{\mathsf{P}}{\to} 0$$

as $T \to +\infty$. Thus, the processes $\beta_T^{(1)}(t)$ converge weakly, as $T \to +\infty$, to the process $2b_0 W(t)$. □

6.4 Weak Convergence of Martingale Type Functional and of Mixed Functional

In this section we formulate sufficient conditions for the weak convergence of some integral functional of martingale type and of the mixed functional.

Theorem 6.6 *Let ξ_T be a solution of Eq. (6.1) belonging to the class $K(G_T)$ and let the assumptions of Theorem 6.2 hold. Assume that, for measurable and locally bounded functions $g_T(x)$, there exists a measurable locally bounded function $g_0(x)$ such that the function*

$$q_T(x) = \left[g_T(x) - g_0(G_T(x)) G_T'(x)\right]^2$$

satisfies assumption (A_3).

Then the stochastic processes

$$\beta_T^{(2)}(t) = \int_0^t g_T(\xi_T(s)) \, dW_T(s),$$

where ξ_T and W_T related via Eq. (5.1), converge weakly, as $T \to +\infty$, to the process

$$\beta^{(2)}(t) = \int_0^t g_0(\zeta(s)) \sigma_0(\zeta(s)) \, d\widehat{W}(s),$$

where $\left(\zeta, \widehat{W}\right)$ is the solution of Eq. (5.14).

The proof of Theorem 6.6 is literally the same as that of Theorem 5.6.

Theorem 6.7 *Let ξ_T and W_T be related via Eq. (6.1) belonging to the class $K(G_T)$ and let the assumptions of Theorem 6.2 hold. Assume that, for continuous functions F_T and locally bounded measurable functions g_T, there exist a continuous function F_0 and a locally bounded measurable function g_0 such that, for all $N > 0$*

$$\lim_{T \to +\infty} \sup_{|x| \leq N} |F_T(x) - F_0(G_T(x))| = 0.$$

Also, let the functions g_T and g_0 satisfy the assumptions of Theorem 6.6.
   Then the stochastic processes

$$I_T(t) = F_T(\xi_T(t)) + \int_0^t g_T(\xi_T(s)) \, dW_T(s)$$

converge weakly, as $T \to +\infty$, to the process

$$I_0(t) = F_0(\zeta(t)) + \int_0^t g_0(\zeta(s)) \sigma_0(\zeta(s)) \, d\widehat{W}(s),$$

where $(\zeta, \widehat{W})$ is the solution of Eq. (5.14).

The proof of Theorem 6.7 is literally the same as that of Theorem 5.9.

6.5   Examples

We denote by b_T a family of constants such that $b_T > 1$ and $b_T \uparrow +\infty$, as $T \to +\infty$.

Example 6.1 Consider Eq. (6.1) with the drift coefficient having non-regular dependence on the parameter T:

$$a_T(t, x) = b_T^\gamma \cos(x b_T) + \frac{t b_T}{1 + t^2 b_T^2} \sin((x - 1) b_T), \quad 0 \leq \gamma < 1.$$

The family of measurable locally bounded real-valued functions $\widehat{a}_T(x) = b_T^\gamma \cos(x b_T)$ satisfies condition (1) from Definition 6.1.
   Indeed, for any $L > 0$

$$\lim_{T \to +\infty} \int_0^L \sup_x |a_T(t, x) - \widehat{a}_T(x)| \, dt \leq \lim_{T \to +\infty} \int_0^L \frac{t b_T}{1 + t^2 b_T^2} \, dt = 0.$$

The rest of the conditions from Definition 6.1 are fulfilled, if we take the family of functions

$$G_T(x) = f_T(x) = \int\limits_0^x \exp\left\{-2\int\limits_0^u \widehat{a}_T(v)\,dv\right\}\,du, \quad T > 0.$$

Since $f_T'(x) = \exp\left\{-2\frac{b_T^\gamma}{b_T}\sin(xb_T)\right\}$, there exist two constants c_0 and δ_0 such that, for all $x \in \mathbb{R}$, we have $0 < \delta_0 \le f_T'(x) \le c_0$. Taking into account that $G_T(x) = \int\limits_0^x f_T'(v)\,dv$, we obtain $G_T'(x)\widehat{a}_T(x) + \frac{1}{2}G_T''(x) \equiv 0$.

Therefore, condition (2) holds, because

$$\left[G_T'(x)a_T(t,x) + \frac{1}{2}G_T''(x)\right]^2 + \left[G_T'(x)\right]^2$$

$$= \left[G_T'(x)\frac{tb_T}{1+t^2b_T^2}\sin((x-1)b_T)\right]^2 + \left[G_T'(x)\right]^2 \le 2\left[G_T'(x)\right]^2$$

$$\le 2c_0^2 \le 2c_0^2\left[1 + |G_T(x)|^2\right];$$

$$|G_T(x_0)| = \left|\int\limits_0^{x_0} f_T'(v)\,dv\right| \le c_0 \cdot |x_0| = C.$$

Condition (3) has the form

$$|G_T(x)| = \left|\int\limits_0^x f_T'(v)\,dv\right| \ge C|x|^\alpha \text{ with } C = \delta_0, \alpha = 1;$$

and condition (4)

$$\left|\int\limits_0^x f_T'(u)\left(\int\limits_0^u \frac{\chi_B(G_T(v))}{f_T'(v)}\,dv\right)du\right| \le \frac{c_0}{\delta_0}\left|\int\limits_0^x\int\limits_0^u \chi_B(G_T(v))\,dv\,du\right|$$

$$\le C_1\lambda(B)|x| \le \psi(\lambda(B))\left[1 + |x|^m\right]$$

is fulfilled with $\psi(|x|) = C_1|x|, m = 1$.

Thus, Eq. (6.1) belongs to the class $K(G_T)$. According to Theorem 6.1, the family of processes $\zeta_T(t) = G_T(\xi_T(t))$ is weakly compact. We can find the form of the limit process using Theorem 6.2 with $a_0(x) \equiv 0$, $\sigma_0(x) \equiv 1$. According to Theorem 6.2, the stochastic processes $\zeta_T(t)$ converge weakly, as $T \to +\infty$, to the solution ζ of Eq. (5.14) and the limit process is $\zeta(t) = x_0 + \widehat{W}(t)$, where $\widehat{W}$ is a Wiener process.

Example 6.2 Let the conditions of Example 6.1 hold. For the family of functions

$$g_T(x) = \frac{b_T^\gamma}{1 + b_T^2 x^2}, \ 0 \le \gamma < 1$$

the assumptions of Theorem 6.3 hold with $g_0(x) \equiv 0$. According to Theorem 6.3, the stochastic processes

$$\beta_T^{(1)}(t) = \int_0^t g_T(\xi_T(s)) \, ds = \int_0^t \frac{b_T^\gamma}{1 + b_T^2 \xi_T^2(s)} \, ds, \ 0 \le \gamma < 1$$

converge weakly, as $T \to +\infty$, to the process $\beta^{(1)}(t) \equiv 0$.

6.6   Auxiliary Results

Lemma 6.1 *Let ξ_T be a solution of Eq. (6.1) belonging to the class $K(G_T)$. Then, for any $N > 0$, $L > 0$ and any Borel set $B \subset [-N; N]$, there exists a constant C_L such that*

$$\int_0^L P\{G_T(\xi_T(s)) \in B\} \, ds \le C_L \psi(\lambda(B)),$$

where $\lambda(B)$ is the Lebesgue measure of the set B, $\psi(|x|)$ is a certain bounded function satisfying the assumption $\psi(|x|) \to 0$, as $|x| \to 0$.

Proof Consider the function

$$\Phi_T(x) = 2 \int_0^x f_T'(u) \left(\int_0^u \frac{\chi_B(G_T(v))}{f_T'(v)} \, dv \right) du.$$

The function $\Phi_T(x)$ is continuous, the derivative $\Phi_T'(x)$ of this function is continuous and the second derivative $\Phi_T''(x)$ exists a.e. with respect to the Lebesgue measure and is locally bounded. Therefore, we can apply the Itô formula to the process $\Phi_T(\xi_T(t))$, where ξ_T is the solution of Eq. (6.1).

Furthermore,

$$\Phi_T'(x)\widehat{a}_T(x) + \frac{1}{2}\Phi_T''(x) = \chi_B(G_T(x)),$$

a.e. with respect to the Lebesgue measure. Using the Itô formula and the latter equality, we conclude that

$$\int_0^L \chi_B(\zeta_T(s))\,ds = \Phi_T(\xi_T(L)) - \Phi_T(x_0) - \int_0^L \Phi_T'(\xi_T(s))\,dW_T(s) - \alpha_T(L)$$

with probability 1 for all $t \geq 0$, where $\zeta_T(t) = G_T(\xi_T(t))$,

$$\alpha_T(t) = \int_0^t \Phi_T'(\xi_T(s))\,[a_T(s, \xi_T(s)) - \widehat{a}_T(\xi_T(s))]\,ds.$$

Hence, using the properties of stochastic integrals, we obtain that

$$\int_0^L P\{\zeta_T(s) \in B\}\,ds = E\,[\Phi_T(\xi_T(L)) - \Phi_T(x_0)] - E\alpha_T(L). \tag{6.13}$$

According to condition (A_2), inequalities $|G_T(x)| \geq C|x|^\alpha$, $C > 0$, $\alpha > 0$ and (5.9), we have that

$$|E\,[\Phi_T(\xi_T(L)) - \Phi_T(x_0)]| \leq C_L^{(1)}\psi\,(\lambda(B))$$

for a certain constant $C_L^{(1)}$. Condition (A_0) implies that

$$|E\alpha_T(L)| \leq C_L^{(2)}\psi\,(\lambda(B))$$

for a certain constant $C_L^{(2)}$. Here the function $\psi\,(\lambda(B))$ is from condition (A_2). The latter inequalities and equality (6.13) prove Lemma 6.1. $\qquad\square$

Lemma 6.2 *Let ξ_T be a solution of Eq. (6.1) belonging to the class $K(G_T)$. If, for measurable locally bounded functions q_T, condition (A_3) holds, then, for any $L > 0$,*

$$\sup_{0 \leq t \leq L} \left| \int_0^t q_T(\xi_T(s))\,ds \right| \xrightarrow{P} 0,$$

as $T \to +\infty$.

Proof Consider the function

$$\Phi_T(x) = 2 \int\limits_0^x f_T'(u) \left(\int\limits_0^u \frac{q_T(v)}{f_T'(v)} \, dv \right) du.$$

The function $\Phi_T(x)$ and the derivative $\Phi_T'(x)$ of this function are continuous, and the second derivative $\Phi_T''(x)$ exists a.e. with respect to the Lebesgue measure and is locally bounded. Therefore, we can apply the Itô formula to the process $\Phi_T(\xi_T(t))$, where ξ_T is the solution to Eq. (6.1).

Furthermore,

$$\Phi_T'(x)\widehat{a}_T(x) + \frac{1}{2}\,\Phi_T''(x) = q_T(x)$$

a.e. with respect to the Lebesgue measure. Using the latter equality, we conclude that with probability 1, for all $t \geq 0$,

$$\int\limits_0^t q_T(\xi_T(s)) \, ds = \Phi_T(\xi_T(t)) - \Phi_T(x_0) - \int\limits_0^t \Phi_T'(\xi_T(s)) \, dW_T(s) - \alpha_T(t),$$

$$(6.14)$$

where

$$\alpha_T(t) = \int\limits_0^t \Phi_T'(\xi_T(s)) \left[a_T(s, \xi_T(s)) - \widehat{a}_T(\xi_T(s)) \right] ds.$$

For any constants $\varepsilon > 0$, $N > 0$ and $L > 0$, we have

$$\mathsf{P} \left\{ \sup_{0 \leq t \leq L} |\alpha_T(t)| > \varepsilon \right\} \leq P_{NT}$$

$$+ \frac{4}{\varepsilon} \sup_{|x| \leq N} f_T'(x) \left| \int\limits_0^x \frac{q_T(v)}{f_T'(v)} \, dv \right| \int\limits_0^L \sup_x \left[a_T(s, x) - \widehat{a}_T(x) \right] ds,$$

where $P_{NT} = \mathsf{P} \left\{ \sup\limits_{0 \leq t \leq L} |\xi_T(t)| > N \right\}.$

The same arguments as used in the proof of Lemma 5.2 and the assumptions of Lemma 6.2 yield that

$$\sup_{0 \le t \le L} |\alpha_T(t)| \xrightarrow{P} 0,$$

$$\sup_{0 \le t \le L} |\Phi_T(\xi_T(t)) - \Phi_T(x_0)| \xrightarrow{P} 0,$$

$$\sup_{0 \le t \le L} \left| \int_0^t \Phi'_T(\xi_T(s)) \, dW_T(s) \right| \xrightarrow{P} 0,$$

as $T \to +\infty$. Thus, the equality (6.14) implies the statement of Lemma 6.2. $\square$

Appendix A
Selected Facts and Auxiliary Results

Let us consider main definitions and some auxiliary results that are used throughout the book. Section A.1 contains well-known facts from the theory of stochastic processes, construction of stochastic integrals, the generalized Itô formula, some classical results for SDEs, Skorokhod's representation theorem, and Prokhorov's theorem. For the proofs, we recommend [20, 59, 65], and some respective references are given in the text. Section A.2 contains more specific results, including convergence results for stochastic integrals and solutions of SDEs. They are given with proofs. Section A.3 is devoted to the detailed description of the Brownian motion in the bilayer environment, and Sect. A.4 contains a brief description of the regularly varying functions.

A.1 Selected Definitions and Facts for Stochastic Processes and Stochastic Integration

A.1.1 Basic Facts Regarding Stochastic Processes

Let $(\Omega, \mathfrak{F}, \mathsf{P})$ be a probability space. Here Ω is a sample space, i.e., a collection of all possible outcomes or results of the experiment, and $\mathfrak{F}$ is a σ-field; in other words, $(\Omega, \mathfrak{F})$ is a measurable space, and P is a probability measure on $\mathfrak{F}$. Let (S, Σ) be another measurable space with σ-field Σ, and let us consider the functions defined on the space $(\Omega, \mathfrak{F})$ and taking their values in (S, Σ). Recall the notion of random variable.

Definition A.1 A random variable on the probability space $(\Omega, \mathfrak{F})$ with the values in the measurable space (S, Σ) is a measurable map $\Omega \overset{\xi}{\to} S$, i.e., a map for which

© Springer Nature Switzerland AG 2020

G. Kulinich et al., *Asymptotic Analysis of Unstable Solutions of Stochastic Differential Equations*, Bocconi & Springer Series 9,
https://doi.org/10.1007/978-3-030-41291-3

the following condition holds: the pre-image $\xi^{-1}(B)$ of any set $B \in \Sigma$ belongs to $\mathfrak{F}$.

Definition A.2 Stochastic process on the probability space $(\Omega, \mathfrak{F}, \mathsf{P})$, parameterized by the set $\mathbb{T}$ and taking values in the measurable space (S, Σ), is a set of random variables of the form

$$X_t = \{X_t(\omega), t \in \mathbb{T}, \omega \in \Omega\},$$

where $X_t(\omega) : \mathbb{T} \times \Omega \to S$.

Thus, each parameter value $t \in \mathbb{T}$ is associated with the random variable X_t taking its value in S.

Here are the other common designations of stochastic processes:

$$X(t), \ \xi(t), \ \xi_t, \ X = \{X_t, t \in \mathbb{T}\}.$$

If $S = \mathbb{R}$, then the process is called real or real valued. Additionally, we assume in this case that $\Sigma = \mathscr{B}(\mathbb{R})$, i.e., $(S, \Sigma) = (\mathbb{R}, \mathscr{B}(\mathbb{R}))$, where $\mathscr{B}(S)$ is a Borel σ-field on S.

Concerning the parameter set $\mathbb{T}$, as a rule, it is interpreted as a time set. If the time parameter is continuous, then usually either $\mathbb{T} = [a, b]$, or $[a, +\infty)$, or $\mathbb{R}^+$. If the time parameter is discrete, then usually either $\mathbb{T} = \mathbb{N} = 1, 2, 3, \ldots$, or $\mathbb{T} = \mathbb{Z}^+ = \mathbb{N} \cup 0$, or $\mathbb{T} = \mathbb{Z}$. We consider the real-valued parameter $\mathbb{T} \subset \mathbb{R}^+$, so that we can regard the parameter as time, as described above.

A stochastic process $X = \{X_t(\omega), t \in \mathbb{T}, \omega \in \Omega\}$ is a function of two variables, one of them being a time variable $t \in \mathbb{T}$ and the other one is a sample point (elementary event) $\omega \in \Omega$. As mentioned earlier, fixing $t \in \mathbb{T}$, we get a random variable $X_t(\cdot)$. In contrast, fixing $\omega \in \Omega$ and following the values that $X.(\omega)$ takes as the function of parameter $t \in \mathbb{T}$, we get a trajectory (path, sample path) of stochastic process. The trajectory is a function of $t \in \mathbb{T}$ and, for any t, it takes its value in S. Changing the value of ω, we get a set of paths, or trajectories, of stochastic process. Assume that the process is real valued and $\mathbb{T} = [a, b]$ or $\mathbb{R}^+$. If its trajectories are a.s. continuous functions, then X is called continuous stochastic process. If its trajectories are a.s. nondecreasing (nonincreasing) functions, then X is called nondecreasing (nonincreasing) stochastic process.

The σ-algebra generated by a stochastic process X is the smallest σ-algebra containing all the sets of the form

$$\{\omega \in \Omega : X(t_1, \omega) \in A_1, \ldots, X(t_k, \omega) \in A_k\}, A_i \in \Sigma, t_i \in \mathbb{T}, 1 \leq i \leq k, k \geq 1.$$

A.1.1.1   Wiener Process

Definition A.3 A stochastic process $X = \{X_t, t \geq 0\}$ is called a process with independent increments, if for any set of points $0 \leq t_1 < t_2 < \ldots < t_n$, the random variables $X_{t_1}, X_{t_2} - X_{t_1}, \ldots, X_{t_n} - X_{t_{n-1}}$ are mutually independent.

Definition A.4 A real-valued stochastic process $W = \{W_t, t \geq 0\}$ is called a (standard) Wiener process if it satisfies the following three conditions:

1. $W_0 = 0$.
2. The process W has independent increments.
3. Increments $W_t - W_s$ for any $0 \leq s < t$ have the Gaussian distribution with zero mean and variance $t - s$. In other words, $W_t - W_s \sim \mathbb{N}(0, t - s)$.

Remark A.1 The Wiener process is often called Brownian motion.

A.1.1.2   Gaussian Processes

Let $X = \{X_t, t \in \mathbb{T}\}$ be a real-valued stochastic process.

Definition A.5 Stochastic process X is Gaussian if all its finite-dimensional distributions are Gaussian, i.e., for any $m \geq 1$ and any $t_1, \ldots, t_m \in \mathbb{T}$ random vector $(X_{t_1}, \ldots, X_{t_m})$ is Gaussian.

For any Gaussian process, there exists function $\{a(t), t \in \mathbb{T}\}$ and function of two variables $\{R(t, s), (t, s) \in \mathbb{T} \times \mathbb{T}\}$ such that for any $m \geq 1$, any $t_i \in \mathbb{T}, 1 \leq i \leq m$ and any $\lambda_1, \ldots, \lambda_m \in \mathbb{R}$

$$\mathsf{E} \exp \left\{ i \sum_{j=1}^{m} \lambda_j X_{t_j} \right\} = \exp \left\{ i \sum_{j=1}^{m} \lambda_j a(t_j) - \frac{1}{2} \sum_{j,k=1}^{m} R(t_j, t_k) \lambda_j \lambda_k \right\}.$$

Function R has the properties:

1. $R(t, s) = R(s, t)$, $(s, t) \in \mathbb{T} \times \mathbb{T}$;
2. for any $m \geq 1$, any $t_1, \ldots, t_m \in \mathbb{T}$ and any $b_1, \ldots, b_m \in \mathbb{R}$

$$\sum_{j,k=1}^{m} R(t_j, t_k) b_j b_k \geq 0.$$

A.1.1.3   Wiener Process as an Example of a Gaussian Process

Consider a Wiener process $W = \{W_t, t \geq 0\}$ satisfying Definition A.4. Recall that for any $t \geq 0$ $\mathsf{E} W_t = 0$ and the covariance function equals

$$Cov(W_s, W_t) = \mathsf{E} \, W_s W_t = s \wedge t.$$

The function $R(s, t) = s \wedge t$ is nonnegative definite, as any covariance function. With this in mind, consider another definition of a Wiener process.

Definition A.6 Stochastic process $W = \{W_t, t \geq 0\}$ is a Wiener process if it satisfies three assumptions:

1. W is a Gaussian process;
2. $\mathsf{E}\, W_t = 0$, for any $t \geq 0$;
3. $Cov(W_s, W_t) = s \wedge t, \quad s, t \geq 0$.

Definitions A.4 and A.6 of the Wiener process are equivalent.

A.1.2   Notion of Stochastic Basis with Filtration

Consider a probability space $(\Omega, \mathfrak{F}, \mathsf{P})$. Let a family $\{\mathfrak{F}_t, t \geq 0\}$ of σ-fields satisfy the following assumptions that are often called "standard assumptions" or "usual conditions."

(i) For any $0 \leq s < t$

$$\mathfrak{F}_s \subset \mathfrak{F}_t \subset \mathfrak{F}.$$

(ii) For any $t \geq 0$

$$\mathfrak{F}_t = \bigcap_{s > t} \mathfrak{F}_s \quad \text{(continuity "from the right").}$$

(iii) $\mathfrak{F}_0$ contains all the sets from $\mathfrak{F}$ of zero P-measure.

Definition A.7 The family $\{\mathfrak{F}_t, t \geq 0\}$ satisfying assumptions (i)–(iii) is called a flow of σ-fields, or a filtration.

Remark A.2 The notion of filtration reflects the fact that information is increasing in time: the more time passed, the more events we could observe, and the richer is corresponding σ-field. Continuity "from the right" means that each σ-field $\mathfrak{F}_t$ is sufficiently rich to contain all "future sprouts," and condition (iii) means the completeness of all σ-fields.

Sometimes the collection $(\Omega, \mathfrak{F}, \{\mathfrak{F}_t\}_{t \geq 0}, \mathsf{P})$ is called a stochastic basis with filtration.

Definition A.8 Stochastic process $X = \{X_t, t \geq 0\}$ is said to be *adapted* to the filtration $\{\mathfrak{F}_t\}_{t \geq 0}$ or simply X is $\mathfrak{F}$-adapted, if for any $t \geq 0$ X_t is $\mathfrak{F}_t$-measurable.

If we write $\{X_t, \mathfrak{F}_t, t \geq 0\}$, then it means that X is $\mathfrak{F}$-adapted.

Remark A.3 Adaptedness of stochastic process means that for any moment of time the values of the process "agree" with the information available at this moment of time.

Remark A.4 Let $X = \{X_t, t \geq 0\}$ be a real stochastic process. We can define σ-algebra $\mathfrak{F}_t^X$ generated by the process X restricted to the interval $[0, t]$: it is the smallest σ-algebra containing the sets $\{\omega \in \Omega : X(t_1, \omega) \in A_1, \ldots, X(t_k, \omega) \in A_k\}$, $A_i \subset \mathbb{R}$, $A_i \in \mathscr{B}(\mathbb{R})$, $t_i \leq t, 1 \leq i \leq k$. We denote it $\mathfrak{F}_t^X = \sigma\{X_s, s \leq t\}$ and say that $\{\mathfrak{F}_t^X\}_{t \geq 0}$ is a natural filtration generated by process X. Any stochastic process is adapted to its natural filtration. Moreover, if X is adapted to $\{\mathfrak{F}_t\}_{t \geq 0}$, then $\mathfrak{F}_t^X \subset \mathfrak{F}_t$ for $t \geq 0$.

A.1.3   Notion of (Sub-, Super-) Martingale. Elementary Properties. Square-Integrable Martingales. Quadratic Variations and Quadratic Characteristics

Let $\left(\Omega, \mathfrak{F}, \{\mathfrak{F}_t\}_{t \in \mathbb{R}^+}, \mathsf{P}\right)$ be a stochastic basis with filtration.

Definition A.9 A stochastic process $\{X_t, t \in \mathbb{R}^+\}$ is said to be a martingale w.r.t. a filtration $\{\mathfrak{F}_t\}_{t \in \mathbb{R}^+}$ if it satisfies the following three conditions:

(i) For any $t \in \mathbb{R}^+$ the random variable $X_t \in \mathscr{L}_1(\Omega, \mathfrak{F}, \mathsf{P})$ (it means that the process X is integrable on $\mathbb{R}^+$).
(ii) For any $t \in \mathbb{R}^+$ X_t is $\mathfrak{F}_t$-measurable, so the process X is $\{\mathfrak{F}_t\}_{t \in \mathbb{R}^+}$-adapted.
(iii) For any $s, t \in \mathbb{R}^+$ such that $s \leq t$ it holds that $\mathsf{E}(X_t|\mathfrak{F}_s) = X_s$ P-a.s.

If we change in condition (iii) the sign $=$ for $\geq$ and obtain $\mathsf{E}(X_t|\mathfrak{F}_s) \geq X_s$ P-a.s. for any $s \leq t$, we get the definition of a *submartingale*; if $\mathsf{E}(X_t|\mathfrak{F}_s) \leq X_s$ P-a.s. for any $s \leq t, s, t \in \mathbb{T}$, then we have a *supermartingale*. A vector process is called (sub-, super-) martingale if the corresponding property has each of its components. Evidently, any martingale is a (sub-, super) martingale. If X is a submartingale, then $-X$ is a supermartingale and vice versa.

Lemma A.1

(1) *Each (sub-, super-) martingale has the same property w.r.t. its natural filtration.*
(2) *Property (iii) is equivalent to the following one: for any $s \leq t, s, t \in \mathbb{R}^+$*
    $\mathsf{E}(X_t - X_s|\mathfrak{F}_s) = 0 \, (\geq 0, \leq 0 \text{ for (sub-, super-) martingales)}.$

Example A.1 Let $\{X_t, \mathfrak{F}_t, t \geq 0\}$ be an integrable process with independent increments, $\mathsf{E}X_t = a_t$. Then

$$\mathsf{E}(X_t - X_s|\mathfrak{F}_s) = \mathsf{E}X_t - \mathsf{E}X_s = a_t - a_s.$$

Therefore, X is a martingale if $a_t = a$, i.e., is the same for any $t \geq 0$, and X is a sub- (super-) martingale if a_t is increasing (decreasing) in t. In particular, Wiener process W is a martingale w.r.t. a natural filtration.

Definition A.10 Process X is called square-integrable, if for any $t \geq 0$ $\mathsf{E} X_t^2 < \infty$. In particular, martingale X is called square-integrable martingale if for any $t \geq 0$ $\mathsf{E} X_t^2 < \infty$.

Now, let $T > 0$ and

$$\pi_n([0, T]) = \{0 = t_0^n < t_1^n < \cdots t_{k_n}^n = T\}$$

be any sequence of partitions of $[0, T]$.

Definition A.11 Let X be a stochastic process.

(i) If for any $T > 0$ and any sequence $\pi_n([0, T])$ such that

$$|\pi_n([0, T])| = \max_{1 \leq k \leq k_n} |t_k^n - t_{k-1}^n| \to 0, n \to \infty$$

there exists an a.s. limit

$$[X]_T = \lim_{n \to \infty} \sum_{k=1}^{k_n} (X_{t_k^n} - X_{t_{k-1}^n})^2,$$

then this limit is called a quadratic variation and we say that X has a quadratic variation. Process $[X]$ is nondecreasing.

(ii) If for any $T > 0$ and any sequence $\pi_n([0, T])$ such that $|\pi_n([0, T])| \to 0, n \to \infty$ there exists an a.s. limit

$$\langle X \rangle_T = \lim_{n \to \infty} \sum_{k=1}^{k_n} \mathsf{E}\left((X_{t_k^n} - X_{t_{k-1}^n})^2 | \mathfrak{F}_{t_{k-1}^n}\right),$$

then this limit is called a quadratic characteristic and we say that X has a quadratic characteristic. Process $\langle X \rangle$ is nondecreasing.

Theorem A.1 *Let X be a continuous square-integrable martingale w.r.t. a filtration $\{\mathfrak{F}_t\}_{t \in \mathbb{R}^+}$. Then it has both quadratic variation $[X]$ and quadratic characteristics $\langle X \rangle$, they coincide, $[X] = \langle X \rangle$ a.s., they are continuous processes, and $X^2 - \langle X \rangle$ is a continuous martingale.*

A.1.4   Markov Moments and Stopping Times

Let $\left(\Omega, \mathfrak{F}, \{\mathfrak{F}_t\}_{t\in\mathbb{R}^+}, \mathsf{P}\right)$ be a stochastic basis with filtration.

Definition A.12

(1) Random variable $\tau = \tau(\omega) : \ \Omega \to \mathbb{R}^+ \cup \{+\infty\}$ is called Markov moment if for any $t \in \mathbb{R}^+$ the event $\{\omega : \tau(\omega) \le t\} \in \mathfrak{F}_t$.
(2) Markov moment $\tau = \tau(\omega)$ is called a stopping time if $\tau < \infty$ a.s.
(3) The σ-algebra generated by the Markov moment τ is the class of events

$$\mathfrak{F}_\tau = \left\{A \in \mathfrak{F} : A \cap \{\tau \le t\} \in \mathfrak{F}_t, t \in \mathbb{R}^+\right\}.$$

Theorem A.2

(1) *In the case* $\mathbb{T} = \mathbb{R}^+$, *a random variable* $\tau : \Omega \to [0, +\infty]$ *is a Markov moment if and only if, for any* $t \in \mathbb{R}^+$, $\{\tau < t\} \in \mathfrak{F}_t$.
(2) *If* τ *is a Markov moment, then for any nondecreasing* $f : \mathbb{R}^+ \cup \{+\infty\} \to \mathbb{R}^+ \cup \{+\infty\}$ *such that* $f(t) \ge t$ *for any* $t \in \mathbb{R}^+$, $f(\tau)$ *is a Markov moment.*
(3) *Let* σ *and* τ *be Markov moments. Then* $\sigma + \tau$, $\sigma \wedge \tau$, $\sigma \vee \tau$ *are Markov moments.*
(4) *Let* $\{\tau_k, k \ge 1\}$ *be the Markov moments. Then* $\sum_{k=1}^{\infty} \tau_k$, $\sup_{k\ge1} \tau_k$, $\inf_{k\ge1} \tau_k$, $\limsup_{k\to\infty} \tau_k$, $\liminf_{k\to\infty} \tau_k$ *are Markov moments.*

Theorem A.3

(1) *Let* τ *be a Markov moment and let the collection of sets* $\mathfrak{F}_\tau$ *be defined according to Definition A.12, (3). Then* $\mathfrak{F}_\tau$ *is indeed a* σ-algebra *and* τ *is* $\mathfrak{F}_\tau$-measurable *random variable.*
(2) *If* $\mathbb{T} = \mathbb{R}_+$, *then* $A \in \mathfrak{F}_\tau$ *if and only if for any* $t \in \mathbb{T}$, $A \cap \{\tau < t\} \in \mathfrak{F}_t$.
(3) *Let* $\sigma \le \tau$ *be two Markov moments. Then* $\mathfrak{F}_\sigma \subset \mathfrak{F}_\tau$.
(4) *For any two Markov moments* σ *and* τ, $\mathfrak{F}_{\sigma\wedge\tau} = \mathfrak{F}_\sigma \cap \mathfrak{F}_\tau$.
(5) *For any sequence of stopping times* $\{\tau_n, n \ge 1\}$, $\mathfrak{F}_{\inf_{n\ge1}\tau_n} = \bigcap_{n\ge1}\mathfrak{F}_{\tau_n}$.
(6) *Let* σ *and* τ *be two Markov moments. Then the events* $\{\sigma = \tau\}$, $\{\sigma \le \tau\}$, *and* $\{\sigma < \tau\}$ *belong to* $\mathfrak{F}_{\sigma\wedge\tau}$.
(7) *Let* $\left\{X_t, \mathfrak{F}_t, t \in \mathbb{R}^+\right\}$ *be an adapted right-continuous stochastic process and* τ *be a Markov moment. Then* X_τ *is* $\mathfrak{F}_\tau$-measurable.

Theorem A.4 ([17, Chapter 1, § 1, Theorem 1]) *Let* ξ *be a process defined and continuous with probability 1 for* $t \ge 0$, $\xi(0) = 0$. *Let the* σ-algebras $\mathfrak{F}_t$ *be defined for all* $t \ge 0$ *and* $\mathfrak{F}_{t_1} \subset \mathfrak{F}_{t_2}$ *for* $t_1 < t_2$. *If*

(1) $\xi(t)$ *is* $\mathfrak{F}_t$-measurable *for all* $t \ge 0$;
(2) $\mathsf{E}\left[\xi(t+h) - \xi(t)\right]|\mathfrak{F}_t = 0$ *with probability 1 for all* $t \ge 0$ *and* $h > 0$;
(3) $\mathsf{E}\left[\xi(t+h) - \xi(t)\right]^2|\mathfrak{F}_t = h$ *with probability 1 for all* $t \ge 0$ *and* $h > 0$,

then ξ *is a Wiener process.*

A.1.5   Construction of Stochastic Integral w.r.t. a Wiener Process and a Square-Integrable Continuous Martingale

Definition A.13 A real-valued stochastic process $\{X(t), t \in \mathbb{R}^+\}$ is called progressively measurable if for any $t > 0$ and Borel set B

$$\{(s, \omega) \in [0, t] \times \Omega : X(s, \omega) \in B\} \in \mathfrak{F}_t \otimes \mathscr{B}([0, t]),$$

where $\mathscr{B}([0, t])$ is the Borel σ-algebra on $[0, t]$.

Now, for $a, b \in \mathbb{R}^+$, $a < b$, we introduce the class $\mathscr{H}_2([a, b])$ of real-valued processes $\{\xi(t), t \in [a, b]\}$ such that

- ξ is progressively measurable;
- $\|\xi\|^2_{\mathscr{H}_2([a,b])} := \int_a^b \mathsf{E}\xi(t)^2 dt < \infty.$

Let us first consider simple processes of the form

$$\xi(t) = \sum_{k=1}^n \alpha_k \chi[a_k, b_k)(t), \tag{A.1}$$

where $n \geq 1$ is an integer, $a \leq a_k < b_k \leq b$ are some real numbers, and α_k is an $\mathfrak{F}_{a_k}$-measurable square-integrable random variable. Clearly, $\xi \in \mathscr{H}_2([a, b])$. Define Itô integral, or stochastic integral, of ξ with respect to W as

$$\int_a^b \xi(t) dW(t) = \sum_{k=1}^n \alpha_k \big(W(b_k) - W(a_k)\big).$$

For notation simplicity, we will also denote

$$I(\xi, W, [a, b]) = I(\xi, [a, b]) = \int_a^b \xi(t) dW(t).$$

Further, we establish several properties of the Itô integral.

Theorem A.5 *Let ξ, ζ be simple processes in $\mathscr{H}_2([a, b])$. Then the following properties are true.*

1. *$I(\xi + \zeta, [a, b]) = I(\xi, [a, b]) + I(\zeta, [a, b])$.*
2. *For any $c \in \mathbb{R}$ $I(c\xi, [a, b]) = cI(\xi, [a, b])$.*
3. *For any $c \in (a, b)$ $I(\xi, [a, b]) = I(\xi, [a, c]) + I(\xi, [c, b])$.*
4. *$\mathsf{E}I(\xi, [a, b]) = 0$. Moreover, $\{I(\xi, [a, t]), t \in [a, b]\}$ is a martingale.*
5. *$\mathsf{E}I(\xi, [a, b])^2 = \|\xi\|^2_{\mathscr{H}_2([a,b])} = \int_a^b \mathsf{E}\xi(t)^2 dt$.*
6. *$\mathsf{E}\big(I(\xi, [a, b])I(\zeta, [a, b]) \mid \mathfrak{F}_a\big) = \int_a^b \mathsf{E}(\xi(t)\zeta(t) \mid \mathfrak{F}_a) dt$, in particular:*

$$\langle \xi, \zeta \rangle_{\mathscr{H}_2([a,b])} := \mathsf{E}\big(I(\xi, [a, b])I(\zeta, [a, b])\big) = \int_a^b \mathsf{E}(\xi(t)\zeta(t)) dt.$$

Lemma A.2 *Let $\xi \in \mathscr{H}_2([a, b])$. Then, there exists a sequence $\{\xi_n, n \geq 1\}$ of simple processes such that*

$$\|\xi - \xi_n\|_{\mathscr{H}_2([a,b])} \to 0, n \to \infty.$$

With this at hand, the extension is done in a standard manner. Namely, if $\{\xi_n, n \geq 1\}$ is a sequence of simple processes converging in $\mathscr{H}_2([a, b])$ to $\xi \in \mathscr{H}_2([a, b])$, then, due to the isometry property, the sequence $\{I(\xi_n, [a, b]), n \geq 1\}$ is a Cauchy sequence in $\mathscr{L}_2(\Omega)$. Then it has a limit in $\mathscr{L}_2(\Omega)$, which justifies the following definition.

Definition A.14 For $\xi \in \mathscr{H}_2([a, b])$, Itô integral of ξ with respect to Wiener process is the limit

$$I(\xi, [a, b]) = \int_a^b \xi(t) dW(t) = \lim_{n \to \infty} I(\xi_n, [a, b]). \qquad (A.2)$$

in $\mathscr{L}^2(\Omega)$, where $\{\xi_n, n \geq 1\}$ is a sequence of simple processes in $\mathscr{H}_2([a, b])$ such that $\|\xi - \xi_n\|_{\mathscr{H}_2([a,b])} \to 0, n \to \infty$.

The properties of Itô integral defined by (A.3) are essentially the same as for simple functions. For completeness, we give them in full.

Theorem A.6 *Let $\xi, \zeta \in \mathscr{H}_2([a, b])$.*

1. $I(\xi + \zeta, [a, b]) = I(\xi, [a, b]) + I(\zeta, [a, b])$ *almost surely;*
2. *For any $c \in \mathbb{R}$* $I(c\xi, [a, b]) = cI(\xi, [a, b])$ *almost surely;*
3. *For any $c \in (a, b)$* $I(\xi, [a, b]) = I(\xi, [a, c]) + I(\xi, [c, b])$ *almost surely;*
4. $\mathsf{E}I(\xi, [a, b]) = 0$. *Moreover, $\{I(\xi, [a, t]), t \in [a, b]\}$ is a martingale;*
5. $\mathsf{E}I(\xi, [a, b])^2 = \|\xi\|^2_{\mathscr{H}_2([a,b])} = \int_a^b \mathsf{E}\xi(t)^2 dt$ *(Itô isometry). Moreover, the process*

$$M(t) = I(\xi, [a, t])^2 - \int_a^t \xi(s)^2 ds, \ t \in [a, b],$$

is a martingale;
6. $\mathsf{E}\big(I(\xi, [a, b])I(\zeta, [a, b])\big) = \langle \xi, \zeta \rangle_{\mathscr{H}_2([a,b])} = \int_a^b \mathsf{E}\xi(t)\zeta(t)dt$.

Theorem A.7 ([17, Chapter 1, § 3, Theorem 2]) *Let $f \in \mathscr{H}_2([a, b])$. Then the separable process $\int_a^t f(s) dW(s), t \in [a, b]$ is continuous with probability 1, and for arbitrary constants $C > 0$ and $N > 0$*

$$\mathsf{P}\left\{ \sup_{a \leq t \leq b} \left| \int_a^t f(s) dW(s) \right| > C \right\} \leq \mathsf{P}\left\{ \int_a^b f^2(t) dt > N \right\} + \frac{N}{C^2}.$$

Theorem A.8 ([17, Chapter 1, § 4, Theorem 3]) *Let the process $f(t)$ be defined for $t \geq 0$ and for all $L > 0$ $f \in \mathscr{H}_2[0, L]$. Assume that*

$$\int\limits_0^{+\infty} f^2(t)\, dt = +\infty$$

with probability 1. Let

$$\tau_t = \inf\{s : \int\limits_0^s f^2(u)\, du \geq t\}.$$

Then the process

$$\zeta_t = \int\limits_0^{\tau_t} f(s)\, dW(s)$$

is a Wiener process.

Remark A.5 The notion of stochastic integral w.r.t. a Wiener process can be, without great difficulties, extended to the notion of stochastic integral w.r.t. a continuous square-integrable martingale η, because the increments of any square-integrable martingale are orthogonal in the sense that for any $0 \leq t_1 < t_2 \leq t_3 < t_4$,

$$\mathsf{E}(\eta_{t_2} - \eta_{t_1})(\eta_{t_4} - \eta_{t_3}) = 0,$$

and only this property (and not the independence of increments) is used in the construction of the stochastic integral. This construction follows the same steps, from simple functions to the functions from the class $\mathscr{H}_2([a, b], \langle \eta \rangle)$ of real-valued processes $\{\xi(t), t \in [a, b]\}$ such that

- ξ is progressively measurable w.r.t. the filtration generated by η;
- $\|\xi\|^2_{\mathscr{H}_2([a,b],\langle\eta\rangle)} := \int_a^b \mathsf{E}\xi(t)^2 d\langle\eta\rangle_t < \infty$.

Recall that $\langle \eta \rangle$ exists, according to Theorem A.1. Denote, for simple function $\xi(t) = \sum_{k=1}^n \alpha_k \chi[a_k, b_k](t)$ from $\mathscr{H}_2([a, b], \langle \eta \rangle)$, integral

$$I(\xi, [a, b], \langle \eta \rangle) = \sum_{k=1}^n \alpha_k \big(\eta(b_k) - \eta(a_k)\big).$$

As before, for any stochastic function from $\mathscr{H}_2([a, b], \langle \eta \rangle)$ there exists a sequence of simple functions from $\mathscr{H}_2([a, b], \langle \eta \rangle)$ such that $\|\xi - \xi_n\|_{\mathscr{H}_2([a,b],\langle\eta\rangle)} \to 0$, $n \to \infty$.

Definition A.15 For $\xi \in \mathscr{H}_2([a, b], \langle \eta \rangle)$, Itô integral of ξ with respect to continuous square-integrable martingale η is the limit

$$I(\xi, [a, b], \langle \eta \rangle) = \int_a^b \xi(t) d\eta(t) = \lim_{n \to \infty} I(\xi_n, [a, b], \langle \eta \rangle) \qquad (A.3)$$

in $\mathscr{L}^2(\Omega)$, where $\{\xi_n, n \geq 1\}$ is a sequence of simple processes in $\mathscr{H}_2([a, b])$ such that $\|\xi - \xi_n\|_{\mathscr{H}_2([a,b], \langle \eta \rangle)} \to 0, n \to \infty$.

Now, Theorem A.6 can be reformulated in the following way.

Theorem A.9 *Let $\xi, \zeta \in \mathscr{H}_2([a, b], \langle \eta \rangle)$. Then*

1. *$I(\xi + \zeta, [a, b], \eta) = I(\xi, [a, b], \eta) + I(\zeta, [a, b], \eta)$ almost surely;*
2. *For any $c \in \mathbb{R}$ $I(c\xi, [a, b], \eta) = cI(\xi, [a, b], \eta)$ almost surely;*
3. *For any $c \in (a, b)$ $I(\xi, [a, b], \eta) = I(\xi, [a, c], \eta) + I(\xi, [c, b], \eta)$ almost surely;*
4. *$\mathsf{E}I(\xi, [a, b], \eta) = 0$. Moreover, $\{I(\xi, [a, t], \eta), t \in [a, b]\}$ is a martingale;*
5. *$\mathsf{E}I(\xi, [a, b])^2 = \|\xi\|^2_{\mathscr{H}_2([a,b], \langle \eta \rangle)} = \int_a^b \mathsf{E}\eta(t)^2 d\langle \eta \rangle_t$. Moreover, the process*

$$M(t) = I(\xi, [a, t], \eta)^2 - \int_a^t \xi(s)^2 d\langle \eta \rangle_s, \ t \in [a, b],$$

   is a martingale;
6. *$\mathsf{E}\big(I(\xi, [a, b], \eta) I(\zeta, [a, b], \eta)\big) = \langle \xi, \zeta \rangle_{\mathscr{H}_2([a,b], \langle \eta \rangle)} = \int_a^b \mathsf{E}\xi(t)\zeta(t) d\langle \eta \rangle_t$.*

A.1.6   Generalized Itô Formula

For the definition and some properties of solutions of the Itô equations see Sect. 2.1, in particular, Definition 2.1.

Lemma A.3 *Let ξ be a solution of the Itô equation*

$$\xi(t) = \xi(0) + \int_0^t a(\xi(s)) \, ds + \int_0^t \sigma(\xi(s)) \, dW(s), \qquad (A.4)$$

where $\xi(0)$ is a random variable not depending on W and let $a(x), \sigma(x), x \in \mathbb{R}$, be real-valued measurable functions satisfying the following assumptions:

(a) *there exists a constant $L > 0$ such that for all $x \in \mathbb{R}$*

$$a^2(x) + \sigma^2(x) \leq L\left(1 + |x|^2\right),$$

(b) *for any constant $N > 0$ and for $|x| \leq N$ we have that $\sigma(x) \geq \delta_N > 0$.*

Assume that a function $\Phi(x)$ has continuous derivative $\Phi'(x)$, and the second derivative $\Phi''(x)$ is assumed to exist a.e. with respect to the Lebesgue measure and to be locally integrable. Then with probability 1, for all $t \geq 0$, the following equality holds (Itô's formula for the process $\Phi(\xi(t))$):

$$\Phi(\xi(t)) = \Phi(\xi(0)) + \int_0^t \left[\Phi'(\xi(s)) a(\xi(s)) + \frac{1}{2}\Phi''(\xi(s)) \sigma^2(\xi(s)) \right] ds$$

$$+ \int_0^t \Phi'(\xi(s)) \sigma(\xi(s)) \, dW(s).$$

This formula follows from the more general result of M.V. Krylov [28, Theorem 4]. Consider the homogeneous SDE

$$d\xi(t) = a(\xi(t)) \, dt + \sigma(\xi(t)) \, dW(t), \tag{A.5}$$

where $a(x) : \mathbb{R} \to \mathbb{R}$ and $\sigma(x) : \mathbb{R} \to \mathbb{R}$ are measurable functions. Assume that the functions $a(x)$, $\sigma(x)$, and $\frac{1}{\sigma(x)}$ are locally bounded. Let $f(x)$ be defined in (A.9) and

$$\varphi(x) = f^{-1}(x), \quad \widehat{\sigma}(x) = f(\varphi(x)) \sigma(\varphi(x)),$$

$$r_1 = \int_0^{-\infty} \exp\left\{ -2 \int_0^z \frac{a(y)}{\sigma^2(y)} \, dy \right\} dz = f(-\infty),$$

$$r_2 = \int_0^{+\infty} \exp\left\{ -2 \int_0^z \frac{a(y)}{\sigma^2(y)} \, dy \right\} dz = f(+\infty),$$

$$0 > r_1 \geq -\infty, \quad 0 < r_2 \leq +\infty.$$

Lemma A.4 ([80, Chapter I, § 3, Lemma 9]) *Let ξ be a solution of Eq. (A.5) on the interval $[0, \tau)$, where τ is some Markov moment. Then the process $\zeta(t) = f(\xi(t))$ is a solution of the equation*

$$d\zeta(t) = \widehat{\sigma}(\zeta(t)) \, dW(t) \tag{A.6}$$

on $[0, \tau)$ and $\zeta \in (r_1, r_2)$ for $s < \tau$.

Theorem A.10 ([80, Chapter I, § 3, Theorem 16]) *Let $f(-\infty) = -\infty$, $f(+\infty) = +\infty$, then the process $\zeta(t) = f(\xi(t))$ is a solution of Eq. (A.6).*

If additionally $\int_{\mathbb{R}} \widehat{\sigma}^{-2}(z)\,dz < \infty$, then the process ζ is ergodic with ergodic distribution

$$F(x) = k \int_{-\infty}^{x} \widehat{\sigma}^{-2}(z)\,dz, \quad k = \left(\int_{\mathbb{R}} \widehat{\sigma}^{-2}(z)\,dz \right)^{-1}.$$

Theorem A.11 (S. Nakao [67]) *Let $a(x)$ and $\sigma(x)$ be bounded Borel functions. Suppose $\sigma(x)$ is of bounded variation on any compact interval. Further, suppose that there exists a constant $\delta > 0$ such that $\sigma(x) \geq \delta$ for $x \in \mathbb{R}$. Then, the pathwise uniqueness holds for (A.5).*

Let Eq. (A.5) have unique solution for any initial conditions $\xi(0)$ and $\sigma(x) > 0$ for all x.

Lemma A.5 ([17, Chapter 1, § 16, Remark 1]) *Let $f(-\infty) = -\infty$ and $f(x) \leq C$ for all x. Then*

$$\mathsf{P}\left\{ \lim_{t \to +\infty} \xi(t) = +\infty \right\} = \mathsf{P}\left\{ \sup_{t>0} \xi(t) = +\infty \right\} = \mathsf{P}\left\{ \inf_{t>0} \xi(t) > -\infty \right\} = 1.$$

Lemma A.6 ([17, Chapter 1, § 16, Lemma 2]) *Let $f(+\infty) = +\infty$ and $f(x) \geq C$ for all x. Then*

$$\mathsf{P}\left\{ \sup_{t>0} \xi(t) < +\infty \right\} = \mathsf{P}\left\{ \inf_{t>0} \xi(t) = -\infty \right\} = \mathsf{P}\left\{ \lim_{t \to +\infty} \xi(t) = -\infty \right\} = 1.$$

A.1.7   Skorokhod's Representation Theorem and Prokhorov's Theorem

The following two lemmas contain the well-known result of A.V. Skorokhod, namely Skorokhod's convergent subsequence principle (see [79, Chapter I, § 6]) and the well-known result of Y.V. Prokhorov [73].

Theorem A.12 (Skorokhod's Representation Theorem) *Let the d-dimensional random processes $\xi_n(t)$, $t \geq 0$, $n \geq 1$ be defined on a certain probability space $(\Omega, \mathfrak{F}, \mathsf{P})$. Assume that for any $L > 0$ and $\varepsilon > 0$*

$$\lim_{N \to +\infty} \limsup_{n \to +\infty} \sup_{0 \leq t \leq L} \mathsf{P}\left\{ |\xi_n(t)| > N \right\} = 0,$$

$$\lim_{h \to 0} \limsup_{n \to +\infty} \sup_{|t_1 - t_2| \leq h, \; t_i \leq L} \mathsf{P}\left\{ |\xi_n(t_1) - \xi_n(t_2)| > \varepsilon \right\} = 0.$$

Then we can choose a subsequence $n_k \to +\infty$, a probability space $(\widetilde{\Omega}, \widetilde{\mathfrak{F}}, \widetilde{\mathsf{P}})$, and stochastic processes $\widetilde{\xi}_{n_k}(t)$ and $\widetilde{\xi}(t)$ defined on this space such that the finite-dimensional distributions of the processes $\widetilde{\xi}_{n_k}(t)$ coincide with the finite-dimensional distributions of the processes $\xi_{n_k}(t)$ and, moreover, $\widetilde{\xi}_{n_k}^{(i)}(t) \xrightarrow{\widetilde{\mathsf{P}}} \widetilde{\xi}^{(i)}(t)$, $i = \overline{1, d}$, as $n_k \to +\infty$, for all $t \geq 0$.

In what follows the processes $\widetilde{\xi}_{n_k}$ and ξ_{n_k}, satisfying assumptions of Theorem A.12, are called weakly equivalent.

Theorem A.13 (Prokhorov's Theorem) *Let the stochastic processes ζ_n, $n \geq 1$ and ζ be continuous with probability 1 on the interval $[a, b]$. Let the finite-dimensional distributions of the processes ζ_n converge, as $n \to +\infty$, to the corresponding finite-dimensional distributions of the process ζ.*

The stochastic processes ζ_n weakly converge in the uniform topology of the space of continuous functions, as $n \to +\infty$, to the process ζ on the interval $[a, b]$ if and only if for any $\varepsilon > 0$:

$$\lim_{h \to 0} \sup_n \mathsf{P} \left\{ \sup_{|t_1 - t_2| \leq h, \, a \leq t_i \leq b} |\zeta_n(t_1) - \zeta_n(t_2)| > \varepsilon \right\} = 0.$$

A.2   Convergence of Stochastic Integrals and Some Properties of Solutions of SDEs

Consider some conditions of the convergence of stochastic integrals w.r.t. square-integrable continuous martingales.

Lemma A.7 *Let $\{\eta_n, n \geq 1\}$ be a sequence of square-integrable continuous martingales with respect to the σ-algebras $\sigma\{\eta_n(s), s \leq t\}$, g_n be real-valued measurable locally bounded functions, and the stochastic integrals $\int_0^L g_n(\xi_n(s))d\eta_n(s)$ are well defined on the probability space $(\Omega, \mathfrak{F}, \mathsf{P})$ for each $n \geq 0$. Assume that g_n converges, as $n \to +\infty$, to the function g_0, almost everywhere (a.e.) with respect to the Lebesgue measure. If*

(1) $\lim\limits_{N \to +\infty} \limsup\limits_{n \to +\infty} \mathsf{P}\{ \sup\limits_{0 \leq t \leq L} |\xi_n(t)| > N \} = 0$;

(2) $\lim\limits_{h \to 0} \limsup\limits_{n \to +\infty} \sup\limits_{|t_2 - t_1| \leq h} \{ \mathsf{P}|\xi_n(t_2) - \xi_n(t_1)| > \varepsilon \} = 0$ *for every* $\varepsilon > 0$;

(3) $\xi_n(t) \xrightarrow{\mathsf{P}} \xi_0(t)$, $\eta_n(t) \xrightarrow{\mathsf{P}} \eta_0(t)$, *as* $n \to +\infty$, *where* η_0 *is a square-integrable continuous martingale with respect to the σ-algebra $\sigma\{\eta_0(s), s \leq t\}$, and for the quadratic characteristics $\langle \eta_n \rangle$ we have the inequality*

$$| \langle \eta_n \rangle (t_2) - \langle \eta_n \rangle (t_1)| \leq C|t_2 - t_1| + |\lambda_n(t_2) - \lambda_n(t_1)|$$

with $E \overset{L}{\underset{0}{V}} \lambda_n \to 0$, *as* $n \to +\infty$, *where* $\overset{L}{\underset{0}{V}} \lambda_n$ *is the variation of the (possibly, random) functions* λ_n *on an interval* $[0, L]$;

(4) $\limsup\limits_{n \to +\infty} \int_0^L P\{\xi_n(s) \in A\} \, ds \leq C\lambda(A)$, *where* A *an arbitrary bounded measurable set,* $C = C(L)$, $\lambda(A)$ *is the Lebesgue measure of the set A.*

Then

$$\int_0^T g_n(\xi_n(s)) \, d\eta_n(s) \overset{P}{\longrightarrow} \int_0^T g_0(\xi_0(s)) \, d\eta_0(s), \qquad (A.7)$$

as $n \to +\infty$.

Proof For an arbitrary constants $N > 0$ and $\delta > 0$ there exists a subset $A_\delta \subset [-N, N]$ with $\lambda(A_\delta) < \delta$ such that according to Luzin's theorem (see [78]), there exists a continuous function $\widetilde{g}(x)$, $x \in [-N, N]$ that coincides with $g_0(x)$ for $x \notin (A_\delta)$. According to Egorov's theorem (see [78]), the pointwise convergence almost everywhere on $[-N, N]$ of the sequence of functions $g_n(x)$ to the function $g(x)$ implies uniform convergence to the function $\widetilde{g}(x)$ everywhere except on the set $A_\delta \subset [-N, N]$.

Let $q_N(x)$ be a continuous nonnegative function that equals to 1, for $|x| \leq N$, and equals to 0, for $|x| > N + 1$.

Then

$$\int_0^L g_n(\xi_n(s)) \, d\eta_n(s) = \int_0^L g_n(\xi_n(s)) q_N(\xi_n(s)) \chi_{A_\delta^c}(\xi_n(s)) \, d\eta_n(s) + \alpha_N + \alpha_{N,\delta},$$

where

$$\alpha_N = \int_0^L g_n(\xi_n(s)) \left[1 - q_N(\xi_n(s))\right] \, d\eta_n(s),$$

$$\alpha_{N,\delta} = \int_0^L g_n(\xi_n(s)) q_N(\xi_n(s)) \chi_{A_\delta}(\xi_n(s)) \, d\eta_n(s),$$

$\chi_A(x)$ is the indicator function of a set A, $A_\delta^c = [-N, N] \setminus A_\delta$. Let

$$\sup_{|x| \leq N} |g(x)| \leq C_N.$$

Since

$$P\{|\alpha_N| > \varepsilon\} \leq P\left\{\sup_{0 \leq t \leq L} |\xi_n(t)| > N\right\} \text{ for any } \varepsilon > 0, \text{ and}$$

$$E|\alpha_{N,\delta}|^2 \leq E\int_0^L g_n^2(\xi_n(s))q_N^2(\xi_n(s))\chi_{A_\delta}(\xi_n(s))\, d\langle\eta_n\rangle(s)$$

$$\leq C_N^2 C \int_0^L P\{\xi_n(s) \in A_\delta\}\, ds + C_N^2 E \overset{L}{\underset{0}{V}} \lambda_n,$$

then, according to the conditions of our lemma, we have

$$\int_0^L g_n(\xi_n(s))\, d\eta_n(s) = \int_0^L g_n(\xi_n(s))q_N(\xi_n(s))\chi_{A_\delta^c}(\xi_n(s))\, d\eta_n(s) + o(1),$$

where

$$\lim_{N \to +\infty} \lim_{\delta \to 0} \lim_{n \to +\infty} o(1) = 0$$

in probability.

Note that

$$\int_0^L g_n(\xi_n(s))q_N(\xi_n(s))\chi_{A_\delta^c}(\xi_n(s))\, d\eta_n(s) = \int_0^L \tilde{g}(\xi_n(s))q_N(\xi_n(s))\, d\eta_n(s) + \beta_n^{(1)} + \beta_n^{(2)},$$

where

$$\beta_n^{(1)} = \int_0^L \tilde{g}(\xi_n(s))q_N(\xi_n(s))\chi_{A_\delta}(\xi_n(s))\, d\eta_n(s),$$

and

$$\beta_n^{(2)} = \int_0^L [g_n(\xi_n(s)) - \tilde{g}(\xi_n(s))]\, q_N(\xi_n(s))\chi_{A_\delta^c}(\xi_n(s))\, d\eta_n(s).$$

Moreover,

$$E\left[\beta_n^{(1)}\right] = E \int_0^L \left[\widetilde{g}(\xi_n(s))q_N(\xi_n(s))\right]^2 \chi_{A_\delta}(\xi_n(s)) \, d\langle\eta_n\rangle(s)$$

$$\leq C_N^2 \int_0^L P\{\xi_n(s) \in A_\delta\} \, ds + C_N^2 C E \overset{L}{\underset{0}{V}} \lambda_n,$$

and

$$E\left[\beta_n^{(2)}\right] = E \int_0^L [g_n(\xi_n(s)) - \widetilde{g}(\xi_n(s))]^2 \, q_N^2(\xi_n(s))\chi_{A_\delta^c}(\xi_n(s)) \, d\langle\eta_n\rangle(s)$$

$$\leq \sup_{x \in A_\delta^c} |g_n(x) - \widetilde{g}(x)|^2 E \int_0^L q_N^2(\xi_n(s)) d\langle\eta_n\rangle(s).$$

Therefore, using uniform convergence, as $n \to +\infty$ of the sequence of functions g_n to the function $\widetilde{g}$ on the set A_δ^c, we obtain

$$\int_0^L g_n(\xi_n(s)) \, d\eta_n(s) = \int_0^L \widetilde{g}(\xi_n(s))q_N(\xi_n(s)) \, d\eta_n(s) + o(1), \tag{A.8}$$

as $n \to +\infty$.

Furthermore,

$$P\left\{\left|\int_0^L \widetilde{g}(\xi_n(s))q_N(\xi_n(s)) \, d\eta_n(s) - \int_0^L \widetilde{g}(\xi_n(s)) \, d\eta_n(s)\right| > 0\right\} \leq P\left\{\sup_{0\leq t\leq L} |\xi_n(t)| > N\right\},$$

and

$$P\left\{\left|\int_0^L \widetilde{g}(\xi_0(s))q_N(\xi_0(s)) \, d\eta_0(s) - \int_0^L \widetilde{g}(\xi_0(s)) \, d\eta_0(s)\right| > 0\right\}$$

$$\leq P\left\{\sup_{0\leq t\leq L} |\xi_0(t)| > N\right\} \leq \limsup_{n\to+\infty} P\{|\xi_n(t)| > N\}.$$

Therefore, to prove the lemma, it is sufficient to show that for all $N > 0$

$$\int_0^L \widetilde{g}(\xi_n(s))q_N(\xi_n(s))\,d\eta_n(s) \xrightarrow{\text{P}} \int_0^L \widetilde{g}(\xi_0(s))q_N(\xi_0(s))\,d\eta_0(s),$$

as $n \to +\infty$. Let $0 = s_0 < s_1 < \ldots < s_m = L$ be an arbitrary partition of the interval $[0, L]$. Then

$$\int_0^L \widetilde{g}(\xi_n(s))\,q_N(\xi_n(s))\,d\eta_n(s) - \int_0^L \widetilde{g}(\xi_0(s))q_N(\xi_0(s))\,d\eta_0(s)$$

$$= \sum_{k=0}^{m-1} \widetilde{g}(\xi_n(s_k))q_N(\xi_n(s_k))[\eta_n(s_{k+1}) - \eta_n(s_k)]$$

$$- \sum_{k=0}^{m-1} \widetilde{g}(\xi_0(s_k))q_N(\xi_0(s_k))[\eta_0(s_{k+1}) - \eta_0(s_k)]$$

$$+ \sum_{k=0}^{m-1} \int_{s_k}^{s_{k+1}} [\widetilde{g}(\xi_n(s))q_N(\xi_n(s)) - \widetilde{g}(\xi_n(s_k))q_N(\xi_n(s_k))]\,d\eta_n(s)$$

$$+ \sum_{k=0}^{m-1} \int_{s_k}^{s_{k+1}} [\widetilde{g}(\xi_0(s))q_N(\xi_0(s)) - \widetilde{g}(\xi_0(s_k))q_N(\xi_0(s_k))]\,d\eta_0(s)$$

$$= I_n^{(1)} + I_n^{(2)} + I_n^{(3)} + I_n^{(4)}.$$

According to the convergence $\xi_n(t) \xrightarrow{\text{P}} \xi_0(t)$ and $\eta_n(t) \xrightarrow{\text{P}} \eta_0(t)$, as $n \to +\infty$, we have $I_n^{(1)} + I_n^{(2)} \xrightarrow{\text{P}} 0$, as $n \to +\infty$.

It is clear that

$$\mathsf{E}\left|I_n^{(3)}\right|^2 = \sum_{k=0}^{m-1} \int_{s_k}^{s_{k+1}} [\widetilde{g}(\xi_n(s))q_N(\xi_n(s)) - \widetilde{g}(\xi_n(s_k))q_N(\xi_n(s_k))]^2\,d\langle\eta_n\rangle(s)$$

$$\leq CL \sup_{|t_2-t_1|\leq h} \mathsf{E}\,|\widetilde{g}(\xi_n(t_2))q_N(\xi_n(t_2)) - \widetilde{g}(\xi_n(t_1))q_N(\xi_n(t_1))|^2 + 4C_N^2 \mathsf{E}\,\overset{L}{\underset{0}{V}}\,\lambda_n,$$

where $h = \max_{0 \leq k \leq m-1}[s_{k+1} - s_k]$.

Since the function $\widetilde{g}(x)\,q_N(x)$ is continuous on the set $|x| \leq N+1$, it is uniformly continuous. Therefore, for an arbitrary $\varepsilon > 0$ there exists $\delta_1 > 0$ such that

$$\sup_{|t_2-t_1|\leq h} \mathsf{E}|\widetilde{g}(\xi_n(t_2))q_N(\xi_n(t_2)) - \widetilde{g}(\xi_n(t_1))q_N(\xi_n(t_1))|^2$$

$$\leq \varepsilon^2 + 4C_N^2 \sup_{|t_2-t_1|\leq h} \mathsf{P}\{|\xi_n(t_2) - \xi_n(t_1)| > \delta_1\}.$$

Hence, from the previous inequalities we have that

$$\lim_{h\to 0}\ \lim_{n\to+\infty}\ I_n^{(3)} = 0$$

in probability. Similarly we obtain that

$$\lim_{h\to 0}\ \lim_{n\to+\infty}\ I_n^{(4)} = 0$$

in probability.

Thus, we have the convergence

$$\int_0^L \widetilde{g}(\xi_n(s))q_N(\xi_n(s))\,d\eta_n(s) \xrightarrow{\mathsf{P}} \int_0^t \widetilde{g}(\xi_0(s))q_N(\xi_0(s))\,d\eta_0(s),$$

as $n \to +\infty$ for any $\delta > 0$ and $N > 0$.

The proof follows from the representation (A.8), combined with the equality $\widetilde{g}(x) = g_0(x)$, as $x \notin A_\delta$, and condition (4) from the assumptions. $\square$

Corollary A.1 *If the processes ξ_n and η_n satisfy the conditions (1)–(3) of Lemma A.7, and the function g_0 is continuous, then we have convergence (A.7). In particular, for $g_n(x) = x$, $n \geq 0$, $\eta_0(t) = W(t)$, where W is a Wiener process, such a convergence was obtained in [79, Chapter II, § 3].*

Lemma A.8 *Let ξ be a solution of the Itô equation (A.4), $\xi(0) = x_0$ and $f(-\infty) = -\infty$, $f(+\infty) = +\infty$, where*

$$f(x) = \int_0^x \exp\left\{-2\int_0^u \frac{a(v)}{\sigma^2(v)}\,dv\right\} du. \tag{A.9}$$

Also, let g be a locally square-integrable real-valued function such that $g^2(x) > 0$ for $x \in A$ with a positive Lebesgue measure $\lambda(A) > 0$. Then with probability 1

$$\int_0^{+\infty} g^2(\xi(s))\,ds = +\infty.$$

Proof We may assume without loss of generality that

$$\inf_{-1<x<0} g^2(x) > 0.$$

Let $\sigma_1 = \min\{t : \xi(t) = 0\}$, $\tau_1 = \min\{t > \sigma_1 : \xi(t) = -1\}, \ldots, \sigma_n = \min\{t > \tau_{n-1} : \xi(t) = 0\}$, $\tau_n = \min\{t > \sigma_n : \xi(t) = -1\}, \ldots$, and

$$\zeta_n = \int_{\sigma_n}^{\tau_n} g^2\left(\xi(s)\right) ds.$$

Note that (see [17, Chapter 4, § 16, Lemma 1])

$$P\left\{\limsup_{t\to+\infty} \xi(t) = +\infty\right\} = P\left\{\liminf_{t\to+\infty} \xi(t) = -\infty\right\} = 1.$$

Taking into account that the process ξ has a strong Markov property, it is easy to see that for all $n \geq 1$, $\sigma_n < \tau_n < +\infty$ with probability 1. Moreover, ζ_n is a sequence of independent identically distributed random variables (see [17, Chapter 3, § 15, Corollary 1]). According to the strong law of large numbers,

$$\frac{1}{n}\sum_{i=1}^{n} \zeta_i \to E\zeta_1, \quad n \to +\infty$$

almost surely.

Since the expectation of ζ_1 exists and is nonnegative,

$$E\zeta_1 = E\int_{\sigma_1}^{\tau_1} g^2\left(\xi(s)\right) ds > \inf_{-1<x<0} g^2(x)E(\tau_1 - \sigma_1) > 0,$$

then $\sum_{i=1}^{n} \zeta_i \to +\infty$, as $n \to +\infty$, a.s.

Thus the inequality

$$\int_{0}^{+\infty} g^2\left(\xi(s)\right) ds \geq \sum_{i=1}^{+\infty} \zeta_i$$

implies the statement of the lemma.   □

Lemma A.9 *Let for each $n \geq 1$ $\eta_n(t), t \geq 0$ be martingales on the same probability space and with respect to the σ-algebras $\sigma\{\eta_n(s), s \leq t\}$, $\eta_n(0) = 0$ and $E|\eta_n(t)|^{2+\delta} \leq C$ for some $\delta > 0$. Let $\eta_n(t) \xrightarrow{P} \eta(t)$ and the quadratic* *characteristic* $\langle \eta_n \rangle (t) \xrightarrow{\text{P}} \beta(t)$, *as* $n \to +\infty$, *for any* $t > 0$, *then the limit process* η *is a martingale with respect to the* σ-*algebra* $\sigma \{\eta(s), \ s \le t\}$ *with the quadratic characteristic* $\beta(t)$.

Proof Since $\eta_n(t)$ for each n is a square-integrable martingale, for any $0 \le t_1 < t_2 < \ldots < t_k < t$ and arbitrary bounded function $g(x_1, x_2, \ldots x_k)$ we have

$$\mathsf{E}\eta_n(t)g\,(\eta_n(t_1), \ldots \eta_n(t_k)) = \mathsf{E}\eta_n(t_k)g\,(\eta_n(t_1), \ldots \eta_n(t_k)),$$

$$\mathsf{E}\,[\eta_n(t) - \eta_n(t_k)]^2\,g\,(\eta_n(t_1), \ldots \eta_n(t_k))$$

$$= \mathsf{E}\,(\langle \eta_n \rangle\,(t) - \langle \eta_n \rangle\,(t_k))\,g\,(\eta_n(t_1), \ldots \eta_n(t_k)).$$

Taking into account the inequality $\mathsf{E}\,(\langle \eta_n \rangle(t))^{1+\delta/2} \le C_\delta \mathsf{E}\,|\eta_n(t)|^{2+\delta}$ (see [69]), we pass to the limit, as $n \to +\infty$, in the previous equality and obtain

$$\mathsf{E}\eta(t)g\,(\eta(t_1), \ldots, \eta(t_k)) = \mathsf{E}\eta(t_k)g\,(\eta(t_1), \ldots, \eta(t_k)),$$

$$\mathsf{E}\,[\eta(t) - \eta(t_k)]^2\,g\,(\eta(t_1), \ldots, \eta(t_k))$$

$$= \mathsf{E}\,(\beta(t) - \beta(t_k))\,g\,(\eta(t_1), \ldots, \eta(t_k)).$$

Since the function $g(x_1, \ldots, x_k)$ is arbitrary, we have that $\eta(t)$ is a martingale with respect to $\sigma \{\eta(s), s \le t\}$ with the quadratic characteristic $\langle \eta \rangle(t) = \beta(t)$. □

Lemma A.10 *Let the process* ζ *be the solution of the equation*

$$\zeta(t) = x_0 + \int_0^t \sigma\,(\zeta(s))\,dW(s), \tag{A.10}$$

where σ *is real-valued measurable function and such that* $\sigma^2(x) \le L\left(1 + |x|^2\right)$ *for all* $x \in \mathbb{R}$ *and* $\sigma(x) \ge \delta_N > 0$, *for* $|x| \le N$, *for any* $N > 0$; W *is a Wiener processes defined on a complete probability space* $(\Omega, \mathfrak{F}, \mathsf{P})$. *Let* $x_0 \in (-N, N)$, $\tau_N = \inf\{t : \zeta(t) \notin (-N, N)\}$, g *be a real-valued measurable function for which*

$$\int_{-N}^{N} |g(x)|\,dx < +\infty.$$

Then the inequality holds

$$\mathsf{E} \int_0^{t \wedge \tau_N} |g\,(\zeta(s))|\,ds \le \frac{1}{\delta_N^2} C\,(x_0, L, t) \int_{-N}^{N} |g(x)|\,dx,$$

where $C\,(x_0, L, t)$ *does not depend on* N.

Proof Consider the function

$$Q(x) = 2 \int_0^x \left(\int_0^u |g(v)| \sigma^{-2}(v) \, dv \right) du.$$

According to the Itô formula (see Lemma A.3) we obtain

$$Q(\zeta(t \wedge \tau_N)) - Q(x_0)$$

$$= \int_0^{t \wedge \tau_N} Q'(\zeta(s)) \sigma(\zeta(s)) \, dW(s) + \frac{1}{2} \int_0^{t \wedge \tau_N} Q''(\zeta(s)) \sigma^2(\zeta(s)) \, ds.$$

Since τ_N is a Markov moment, taking into account the properties of stochastic integrals [17, §4], we get from the previous equality that

$$\mathsf{E} \int_0^{t \wedge \tau_N} |g(\zeta(s))| \, ds = \mathsf{E}[Q(\zeta(t \wedge \tau_N)) - Q(x_0)]$$

$$\leq \frac{1}{\delta_N^2} \int_{-N}^N |g(x)| \, dx \, \mathsf{E} \, |\zeta(t \wedge \tau_N) - x_0|$$

$$\leq \frac{1}{\delta_N^2} \int_{-N}^N |g(x)| \, dx \left(\mathsf{E} \int_0^{t \wedge \tau_N} \sigma^2(\zeta(s)) \, ds \right)^{\frac{1}{2}}$$

$$\frac{1}{\delta_N^2} \int_{-N}^N |g(x)| \, dx \cdot L \left(\int_0^t \left[1 + \mathsf{E} \, |\zeta(s)|^2 \right] ds \right)^{\frac{1}{2}}.$$

From this inequality and the well-known bound for $\mathsf{E} \, |\zeta(s)|^2$ (see [17, §6]) the statement of Lemma A.10 follows.   □

Lemma A.11 *Let the process ζ be the solution of Eq. (A.10) and let A be a measurable bounded set, $A \subset (-N, N)$. Then*

$$\int_0^t \mathsf{P}\{\zeta(s) \in A\} \, ds \leq \frac{1}{\delta_N^2} C(x_0, L, t) \lambda(A),$$

where $\lambda(\cdot)$ is the Lebesgue measure, and $C(x_0, L, t)$ is the same constant as in Lemma A.10.

Proof Let

$$Q(x) = 2 \int_0^x \left(\int_0^u \chi_A(v) \sigma^{-2}(v) \, dv \right) du.$$

According to the Itô formula we have for $t > 0$

$$Q(\zeta(t)) - Q(x_0)$$

$$= \int_0^t Q'(\zeta(s)) \sigma(\zeta(s)) \, dW(s) + \frac{1}{2} \int_0^t Q''(\zeta(s)) \sigma^2(\zeta(s)) \, ds.$$

From the last equality, in the same way as in the proof of Lemma A.10, we obtain the proof of Lemma A.11.   □

Corollary A.2 *Let the assumptions of Lemma A.11 hold and additionally for all $x \in \mathbb{R}$ and for some constant C_0 let $\sigma(x) \le C_0$. Then*

$$\int_0^t P\{\zeta(s) \in A\} \, ds \le \frac{C_0 \sqrt{t}}{\delta_N^2} \lambda(A).$$

In fact, this statement follows directly from the proof of Lemma A.11, combined with the inequalities

$$E\,|\zeta(t) - x_0| \le \left(E\,|\zeta(t) - x_0|^2 \right)^{\frac{1}{2}} \le \left(\int_0^t E\sigma^2(\zeta(s)) \, ds \right)^{\frac{1}{2}} \le C_0 \sqrt{t}.$$

Lemma A.12 *Let the process ζ be the solution of Eq. (A.10). If for all $t > 0$ the following equality holds a.s.:*

$$F(\zeta(t)) - F(x_0) = \int_0^t g(\zeta(s)) \, d\zeta(s), \tag{A.11}$$

where F is a continuous function, and the function g is locally square-integrable, then for any $x \ne x_0$

$$\frac{F(x) - F(x_0)}{x - x_0} = g(x) = b$$

a.e. with respect to the Lebesgue measure, where b is some constant.

Proof Let $\tau = \inf\{t : \zeta(t) \notin (x_0 - h_1, x_0 + h_2)\}$, $h_i > 0$. It is clear that τ is a Markov moment. As in Theorem 2, §15, [17], we find that $\mathsf{E}\tau = v(x_0)$, where

$$
v(x) = -2 \int\limits_{x_0-h_1}^{x} \left(\int\limits_{x_0-h_1}^{y} \frac{dz}{\sigma^2(z)} \right) dy + 2 \int\limits_{x_0-h_1}^{x_0+h_2} \left(\int\limits_{x_0-h_1}^{y} \frac{dz}{\sigma^2(z)} \right) dy \frac{x - x_0 + h_1}{h_1 + h_2}.
$$

The function $v(x)$, for $x \in (x_0 - h_1, x_0 + h_2)$, satisfies the equation

$$
\frac{1}{2} v''(x) \sigma^2(x) = -1
$$

with boundary conditions $v(x_0 - h_1) = v(x_0 + h_2) = 0$.

Thus, $\tau < +\infty$ with probability 1. Taking into account (A.10) and (A.11), using the properties of stochastic integrals, we obtain that

$$
\mathsf{E}\zeta(\tau) = x_0, \qquad \mathsf{E}F(\zeta(\tau)) = F(x_0).
$$

Therefore

$$
\mathsf{P}\{\zeta(\tau) = x_0 - h_1\} = \frac{h_2}{h_1 + h_2}
$$

and

$$
\mathsf{P}\{\zeta(\tau) = x_0 + h_2\} = \frac{h_1}{h_1 + h_2}.
$$

Since

$$
\mathsf{E}F(\zeta(\tau)) = F(x_0 - h_1)\mathsf{P}\{\zeta(\tau) = x_0 - h_1\} + F(x_0 + h_2)\mathsf{P}\{\zeta(\tau) = x_0 + h_2\},
$$

for any $h_i > 0$, we have the equality

$$
\frac{F(x_0) - F(x_0 - h_1)}{h_1} = \frac{F(x_0 + h_2) - F(x_0)}{h_2},
$$

from which it follows that $F(x) - F(x_0) = b(x - x_0)$, where b is some constant. According to equality (A.11), we obtain that

$$
\int\limits_{0}^{t} [g(\zeta(s)) - b] \, d\zeta(s) = 0 \tag{A.12}
$$

with probability 1 for all $t \geq 0$. Since $\int_0^t [g(\zeta(s)) - b] \, d\zeta(s)$ is continuous with probability 1 and a square-integrable martingale, its quadratic characteristics

$$\int_0^t [g(\zeta(s)) - b]^2 \sigma^2(\zeta(s)) \, ds$$

with probability 1, for all $t \geq 0$, is equal to zero. Consequently,

$$\int_{-\infty}^{+\infty} [g(x) - b]^2 \sigma^2(x) \, dx = 0,$$

because otherwise the process

$$\int_0^{\tau_t} [g(\zeta(s)) - b] \sigma(\zeta(s)) \, dW(s),$$

where τ_t is the smallest solution of the equation

$$t = \int_0^{\tau_t} [g(\zeta(s)) - b]^2 \sigma^2(\zeta(s)) \, ds,$$

would be (see [17, §4]) a Wiener process, but according to (A.12),

$$\int_0^{\tau_t} [g(\zeta(s)) - b] \sigma(\zeta(s)) \, dW(s) \equiv 0$$

with probability 1.

Thus, $g(x) = b$ a.e. with respect to the Lebesgue measure, where b is some constant. Lemma A.12 is proved.   □

Lemma A.13 *Let the process ξ be the solution of Eq. (A.4) and the process (ξ, W) be weakly equivalent to the process $(\widetilde{\xi}, \widetilde{W})$, where the process $\widetilde{\xi}$ is continuous with probability 1. If, for any $N > 0$ and for any Borel set $A \subset [-N, N]$, the following inequality holds:*

$$\int_0^t \mathsf{P}\{\xi(t) \in A\} \, ds \leq C_N \lambda(A), \tag{A.13}$$

where $\lambda\,(\cdot)$ is the Lebesgue measure, then

$$\tilde{\xi}\,(t) = \tilde{\xi}\,(0) + \int_0^t a\left(\tilde{\xi}\,(s)\right) ds + \int_0^t \sigma\left(\tilde{\xi}\,(s)\right) d\tilde{W}\,(s)\,. \tag{A.14}$$

Proof First, assume that the coefficients $a\,(x)$ and $\sigma\,(x)$ of Eq. (A.4) are continuous functions. Then the processes $a\,(\xi\,(\cdot))$ and $\sigma\,(\xi\,(\cdot))$ are continuous with probability 1 on the interval $[0,\,t]$. Consider the partition of the interval $[0,\,t]$: $0 = t_{n0} < t_{n1} < \ldots < t_{nn} = t$, $\lambda_n = \max\limits_n \Delta t_{nk}$, where $\Delta t_{nk} = t_{nk+1} - t_{nk}$.

Then

$$S_n^{(1)} = \sum_{k=0}^{n-1} a\,(\xi\,(t_{nk}))\,\Delta t_{nk} \xrightarrow{\text{P}} \int_0^t a\,(\xi\,(s))\,ds, \tag{A.15}$$

as $\lambda_n \to 0$. In accordance with the properties of stochastic integrals,

$$S_n^{(2)} = \sum_{k=0}^{n-1} \sigma\,(\xi\,(t_{nk}))\,[W\,(t_{nk+1}) - W\,(t_{nk})] \xrightarrow{\text{P}} \int_0^t \sigma\,(\xi\,(s))\,dW\,(s)\,, \tag{A.16}$$

as $\lambda_n \to 0$. Therefore,

$$\alpha_n := \xi\,(t) - \xi\,(0) - S_n^{(1)} - S_n^{(2)} \xrightarrow{\text{P}} 0,$$

as $\lambda_n \to 0$. The distribution of the random variables α_n coincides with the distribution of $\tilde{\alpha}_n$, where

$$\tilde{\alpha}_n = \tilde{\xi}\,(t) - \tilde{\xi}\,(0) - \tilde{S}_n^{(1)} - \tilde{S}_n^{(2)}, \tag{A.17}$$

$$\tilde{S}_n^{(1)} = \sum_{k=0}^{n-1} a\left(\tilde{\xi}\,(t_{nk})\right)\,\Delta t_{nk}, \quad \tilde{S}_n^{(2)} = \sum_{k=0}^{n-1} \sigma\left(\tilde{\xi}\,(t_{nk})\right)\,\left[\tilde{W}\,(t_{nk+1}) - \tilde{W}\,(t_{nk})\right].$$

From the convergence $\alpha_n \xrightarrow{\text{P}} 0$ it follows that $\tilde{\alpha}_n \xrightarrow{\tilde{\text{P}}} 0$, as $\lambda_n \to 0$. It is clear that the convergences similar to (A.15) and (A.16) hold. Thus, we pass to the limit, as $\lambda_n \to 0$, in (A.17) and obtain (A.14).

Now, let the coefficients $a(x)$ and $\sigma(x)$ be measurable locally bounded functions. For any $N > 0$ we have the equality

$$\xi(t) = \xi(0) + \int_0^t a(\xi(s)) q_N(\xi(s))\, ds +$$

$$\int_0^t \sigma(\xi(s)) q_N(\xi(s))\, dW(s) + \beta_N(t), \qquad (A.18)$$

where

$$\beta_N(t) = \int_0^t a(\xi(s))[1 - q_N(\xi(s))]\, ds + \int_0^t \sigma(\xi(s))[1 - q_N(\xi(s))]\, dW(s),$$

and $q_N(x)$ is the function defined in Lemma A.7.

It is clear that we have the inequality

$$P\{|\beta_N(t)| > 0\} \le P\left\{ \sup_{0 \le s \le t} |\xi(s)| > N \right\}.$$

Since the process ξ is continuous with probability 1, the right-hand side of this inequality tends to 0, as $N \to +\infty$. Thus, $\beta_N(t) \xrightarrow{P} 0$, as $N \to +\infty$, for every $t > 0$. Using Lusin's theorem [78], we conclude that, for any $\varepsilon > 0$, there exist continuous functions $a_N^{(\varepsilon)}(x)$ and $\sigma_N^{(\varepsilon)}(x)$, $x \in \mathbb{R}$, that coincide with $a(x) q_N(x)$ and $\sigma(x) q_N(x)$, respectively, for $x \notin A_N^{(\varepsilon)}$, where $A_N^{(\varepsilon)} \subset [-N-1, N+1]$, and its Lebesgue measure satisfies the inequality $\lambda\left(A_N^{(\varepsilon)}\right) < \varepsilon$.

Therefore, taking into account (A.18) and the condition of Lemma A.10, we have the convergence

$$\lim_{N \to +\infty} \lim_{\varepsilon \to 0} \lim_{\lambda_n \to 0} \left(\xi(t) - \xi(0) - S_n^{(1)} - S_n^{(2)} \right) = 0$$

in probability P, where

$$S_n^{(1)} = \sum_{k=0}^{n-1} a_N^{(\varepsilon)}(\xi(t_{nk}))\, \Delta t_{nk}, \quad S_n^{(2)} = \sum_{k=0}^{n-1} \sigma_N^{(\varepsilon)}(\xi(t_{nk}))[W(t_{nk+1}) - W(t_{nk})].$$

Thus,

$$\lim_{N\to+\infty} \lim_{\varepsilon\to 0} \lim_{\lambda_n\to 0} \left(\widetilde{\xi}(t) - \widetilde{\xi}(0) - \widetilde{S}_n^{(1)} - \widetilde{S}_n^{(2)}\right) = 0$$

in probability $\widetilde{\mathsf{P}}$, where $\widetilde{S}_n^{(i)}$ are the corresponding integral sums of the processes $\widetilde{\xi}$ and $\widetilde{W}$. Continuation of the proof is the same as in the case of continuous functions $a(x)$ and $\sigma(x)$. The analog of the condition (A.13) holds for the process $\widetilde{\xi}$ according to the weak equivalence of the processes ξ and $\widetilde{\xi}$. □

Corollary A.3 *Let g and q be real measurable locally bounded functions, and let the assumptions of Lemma A.13 hold. Then the process*

$$S(t) = \int_0^t g(\xi(s))\, ds + \int_0^t q(\xi(s))\, dW(s)$$

is stochastically equivalent to the process

$$\widetilde{S}(t) = \int_0^t g\left(\widetilde{\xi}(s)\right) ds + \int_0^t q\left(\widetilde{\xi}(s)\right) d\widetilde{W}(s).$$

In fact, this statement follows directly from the proof of Lemma A.13.

Lemma A.14 *If the solution ξ of an Itô SDE is ergodic with a distribution function F, then it is not stochastically unstable.*

Proof Note that for an ergodic solution ξ for any $N > 0$ we have

$$\mathsf{P}\{|\xi(t)| < N\} = \mathsf{P}\{-N < \xi(t) < N\}$$

$$= \mathsf{P}\{\xi(t) < N\} - \mathsf{P}\{\xi(t) \geq -N\} \to F(N) - F(-N+0),$$

as $t \to +\infty$.

Taking into account the equalities $F(-\infty) = 0$ and $F(+\infty) = 1$, we obtain that there exists N_0 for which

$$F(N_0) - F(-N_0 + 0) = a > 0.$$

Therefore,

$$\lim_{t\to+\infty} \mathsf{P}\{|\xi(t)| < N_0\} = a > 0.$$

So, for any $\varepsilon > 0$ there exists $T_\varepsilon > 0$ so that for every $t > T_\varepsilon$ we have the inequality

$$|P\{|\xi(t)| < N_0\} - a| < \varepsilon.$$

From the obvious inequalities

$$\left| \frac{1}{t} \int_0^t P\{|\xi(s)| < N_0\}\, ds - a \right| = \frac{1}{t} \left| \int_0^t (P\{|\xi(s)| < N_0\} - a)\, ds \right| \le$$

$$\frac{1}{t} \int_0^{T_\varepsilon} |P\{|\xi(s)| < N_0\} - a|\, ds + \frac{1}{t} \int_{T_\varepsilon}^t |P\{|\xi(s)| < N_0\} - a|\, ds \le \frac{L_\varepsilon}{t} + \frac{t - T_\varepsilon}{t}\varepsilon$$

we obtain

$$\limsup_{t \to +\infty} \left| \frac{1}{t} \int_0^t P\{|\xi(s)| < N_0\}\, ds - a \right| \le \varepsilon.$$

Since $\varepsilon > 0$ is arbitrary, we conclude that

$$\lim_{t \to +\infty} \frac{1}{t} \int_0^t P\{|\xi(s)| < N_0\}\, ds = a > 0.$$

For $N = N_0$ the convergence from the Definition 1 of the stochastic instability is not satisfied, that is, ξ is not stochastically unstable.   □

A.3   Brownian Motion in a Bilayer Environment

Consider an SDE of the form

$$d\zeta(t) = \bar{\sigma}(\zeta(t))\, dW(t), \quad t > 0, \quad \zeta(0) = 0, \tag{A.19}$$

where W is a Wiener process, $\bar{\sigma}(x) = \sigma_1$ for $x \ge 0$, and $\bar{\sigma}(x) = \sigma_2$ for $x < 0$, both $\sigma_i > 0$, $i = 1, 2$. The specific feature of Eq. (A.19) is a discontinuity of the coefficient $\bar{\sigma}(x)$ at the point $x = 0$. Because of this reason, classical existence-uniqueness conditions cannot be applied. Necessity of the investigation of Eq. (A.19) first emerged in the study of the limit behavior of normalized unstable solutions of Itô SDEs with continuous coefficients. This problem was studied in [29], where the following relation was established for the limit process ζ:

$$\int_0^L P\{\zeta(s) = 0\}\, ds = 0 \tag{A.20}$$

for any $L > 0$. Due to this result, it was proved that the process ζ is a weak solution of Eq. (A.19), and ζ is a Markov process as the limit of the sequence of Markov processes. Besides this, it is proved in the mentioned paper that equality (A.20) allows to apply the first Kolmogorov equation (see [19, Chapter 8, § 1, Theorem 1]) in order to get the explicit form of the transition density $\rho(t, x, y)$ of the Markov process ζ. Finally, it is established that the transition density has the form

$$\rho(t, x, y) = \begin{cases} \dfrac{1}{\sigma_1\sqrt{2\pi t}}\left[e^{-\frac{(y-x)^2}{2\sigma_1^2 t}} - \dfrac{\sigma_1-\sigma_2}{\sigma_1+\sigma_2} e^{-\frac{(y+x)^2}{2\sigma_1^2 t}} \right], & x \geq 0,\, y > 0, \\[2ex] \dfrac{2\sigma_1}{\sigma_1+\sigma_2} \cdot \dfrac{1}{\sigma_2\sqrt{2\pi t}} e^{-\frac{(\sigma_1 y - \sigma_2 x)^2}{2\sigma_1^2 \sigma_2^2 t}}, & x \geq 0,\, y < 0, \\[2ex] \dfrac{2\sigma_2}{\sigma_1+\sigma_2} \cdot \dfrac{1}{\sigma_1\sqrt{2\pi t}} e^{-\frac{(\sigma_2 y - \sigma_1 x)^2}{2\sigma_1^2 \sigma_2^2 t}}, & x \leq 0,\, y > 0, \\[2ex] \dfrac{1}{\sigma_2\sqrt{2\pi t}}\left[e^{-\frac{(y-x)^2}{2\sigma_2^2 t}} - \dfrac{\sigma_2-\sigma_1}{\sigma_1+\sigma_2} e^{-\frac{(y+x)^2}{2\sigma_2^2 t}} \right], & x \leq 0,\, y < 0. \end{cases} \tag{A.21}$$

So, the weak uniqueness of the solution ζ of Eq. (A.19) is established. Moreover, it follows from the explicit form of the density $\rho(t, x, y)$ that the process ζ is a homogeneous in time Markov process whose transition density is continuous in x and discontinuous in y at the point $y = 0$. Continuity in x implies that the process ζ is a Feller Markov process (see [14, Chapter 2, § 1]), consequently, it has a strong Markov property (see [14, Chapter 3, § 3, Theorem 3.10]). Note that $\bar{\sigma}(x) \geq \sigma_0 > 0$ for any $x \in \mathbb{R}$, and the variation $\overset{N}{\underset{-N}{V}} \sigma(x) = |\sigma_1 - \sigma_2|$ for any $N > 0$. Therefore, the solution of Eq. (A.19) is pathwise unique [67], and consequently, it is strong and strongly unique [84].

Note that in the paper [29] the solution η of Eq. (A.19) with $\bar{\sigma}(x) = a > 0$, $x \geq 0$ and $\bar{\sigma}(x) = 1$, $x < 0$, is obtained as a limit of the normalized unstable solutions of some specific SDEs. This process satisfies condition (A.20). The transition density $\rho_\eta(t, x, y)$ has the form (A.21) with $\sigma_1 = a$ and $\sigma_2 = 1$. Obviously, due to the condition (A.20) and the relation $\bar{\sigma}(x) = \bar{\sigma}(\sigma_2 x)$ for $\sigma_2 > 0$, the process $\zeta(t) := \sigma_2\eta(t)$ is a solution of Eq. (A.19) with $\sigma_1 = a\sigma_2$.

It is well known (see [19, Chapter VI, § 5]), that Eq. (A.19) is the mathematical description of the physical Brownian particle moving in a liquid or gaseous environment, in the presence of collisions of this particle with molecules of the environment, with the assumption that these molecules are in a chaotic temperature movement. The intensity of the collision equals $\bar{\sigma}(x)$ at the point x. Let $\zeta = \zeta(t)$ be a projection of the Brownian particle' position on the coordinate axis at the moment $t > 0$. Under these assumptions, during some small interval Δt, we get the approximate equality

$$\zeta(t + \Delta t) - \zeta(t) \simeq \bar{\sigma}(\zeta(t)) \left[W(t + \Delta t) - W(t) \right]. \tag{A.22}$$

Let $S_1 = \{x : x \geq 0\}$, $S_2 = \{x : x < 0\}$, and the intensity of collision equal $\bar{\sigma}(x) = \sigma_1, x \in S_1$, and $\bar{\sigma}(x) = \sigma_2, x \in S_2$. Then

$$\zeta(t + \Delta t) - \zeta(t) \simeq \sigma_1 \left[W(t + \Delta t) - W(t) \right]$$

for $\zeta(t) \in S_1$, while

$$\zeta(t + \Delta t) - \zeta(t) \simeq \sigma_2 \left[W(t + \Delta t) - W(t) \right]$$

for $\zeta(t) \in S_2$.

Therefore in this case the solution ζ of Eq. (A.19) is a process that describes the trajectory of the movement of the Brownian particle in the bilayer environment $S = S_1 \cup S_2$ with the continuous-type border crossing at the point $x = 0$.

Definition A.16 The solution ζ of Eq. (A.19) is called the process of Brownian motion in the bilayer environment $S = S_1 \cup S_2$.

For the reader's convenience, we give here the proof of formula (A.21).

Lemma A.15 *The transition density $\rho(t, x, y)$ of the solution of Eq. (A.19) has the form (A.21).*

Proof Introduce the solution $u(t, x)$ of the Cauchy problem for the parabolic equation

$$-\frac{\partial u(t, x)}{\partial t} = \frac{1}{2} \bar{\sigma}^2(x) \frac{\partial^2 u(t, x)}{\partial x^2},$$

considered on the set $0 \leq t < s, x \in (-\infty, +\infty)$, with conditions of conjugation

$$u(t, +0) = u(t, -0), \quad \frac{\partial u(t, x)}{\partial x}\bigg|_{x=+0} = \frac{\partial u(t, x)}{\partial x}\bigg|_{x=-0},$$

and the initial condition $u(s, x) = q(x)$. The function $q(x)$ is assumed to be twice continuously differentiable and bounded, together with its derivatives $q'(x)$ and $q''(x)$.

Introduce the function $V(t, x) = u(s - t, x)$. Then $V(x, 0) = q(x)$ and for $t > 0$ and $x \in \mathbb{R}$ we have the equation

$$\frac{\partial V(t, x)}{\partial t} = \frac{1}{2}\overline{\sigma}^2(x)\frac{\partial^2 V(t, x)}{\partial x^2}.$$

This equation is convenient to solve by means of the Laplace transformation. So, we multiply both parts of the equation by e^{-pt}, denoting

$$V^*(p, x) = \int_0^{+\infty} e^{-pt} V(t, x)dt.$$

After integrating with respect to t over the interval $[0, +\infty)$ we get the following differential equations:

$$\frac{\partial^2 V^*(p, x)}{\partial x^2} - \frac{2p}{\sigma_1^2} V^*(p, x) = -\frac{2}{\sigma_1^2}q(x), \, x > 0,$$

and

$$\frac{\partial^2 V^*(p, x)}{\partial x^2} - \frac{2p}{\sigma_2^2} V^*(p, x) = -\frac{2}{\sigma_2^2}q(x), \, x < 0.$$

The solution of this couple of equations has the form

$$V^*(p, x) = -\frac{1}{\sigma_1\sqrt{2p}}e^{\frac{\sqrt{2p}}{\sigma_1}x} \int_{+\infty}^{x} e^{-\frac{\sqrt{2p}}{\sigma_1}y}q(y)\,dy$$

$$+\frac{1}{\sigma_1\sqrt{2p}}e^{-\frac{\sqrt{2p}}{\sigma_1}x} \int_{-\infty}^{x} e^{-\frac{\sqrt{2p}}{\sigma_1}y}q(y)\,dy + C_1 e^{\frac{\sqrt{2p}}{\sigma_1}x} + C_2 e^{-\frac{\sqrt{2p}}{\sigma_1}x}$$

for $x > 0$, and

$$V^*(p, x) = -\frac{1}{\sigma_2\sqrt{2p}}e^{\frac{\sqrt{2p}}{\sigma_2}x} \int_{+\infty}^{x} e^{-\frac{\sqrt{2p}}{\sigma_2}y}q(y)\,dy$$

$$+\frac{1}{\sigma_2\sqrt{2p}}e^{-\frac{\sqrt{2p}}{\sigma_2}x} \int_{-\infty}^{x} e^{-\frac{\sqrt{2p}}{\sigma_2}y}q(y)\,dy + C_3 e^{-\frac{\sqrt{2p}}{\sigma_2}x} + C_4 e^{\frac{\sqrt{2p}}{\sigma_2}x}$$

for $x < 0$, where C_i are arbitrary constants.

Since the function $V^*(p, x)$ is bounded, we have that $C_1 = C_3 = 0$. Applying the conditions of conjugation:

$$V^*(p, +0) = V^*(p, -0), \qquad \frac{\partial V^*(p, +0)}{\partial x} = \frac{\partial V^*(p, -0)}{\partial x},$$

we can determine the constants

$$C_2 = \frac{\sigma_1 - \sigma_2}{\sigma_1 + \sigma_2} \frac{1}{\sigma_1 \sqrt{2p}} \int_{+\infty}^{0} e^{-\frac{\sqrt{2p}}{\sigma_1} y} q(y)\, dy$$

$$-\frac{1}{\sigma_1 \sqrt{2p}} \int_{-\infty}^{0} e^{-\frac{\sqrt{2p}}{\sigma_1} y} q(y)\, dy + \frac{2\sigma_1}{\sigma_1 + \sigma_2} \frac{1}{\sigma_2 \sqrt{2p}} \int_{-\infty}^{0} e^{\frac{\sqrt{2p}}{\sigma_2} y} q(y)\, dy,$$

and

$$C_4 = \frac{\sigma_1 - \sigma_2}{\sigma_1 + \sigma_2} \frac{1}{\sigma_2 \sqrt{2p}} \int_{-\infty}^{0} e^{\frac{\sqrt{2p}}{\sigma_2} y} q(y)\, dy$$

$$+\frac{1}{\sigma_2 \sqrt{2p}} \int_{+\infty}^{0} e^{\frac{\sqrt{2p}}{\sigma_2} y} q(y)\, dy - \frac{2\sigma_2}{\sigma_1 + \sigma_2} \frac{1}{\sigma_1 \sqrt{2p}} \int_{+\infty}^{0} e^{-\frac{\sqrt{2p}}{\sigma_2} y} q(y)\, dy.$$

Next, we use the inverse Laplace transform and get

$$V(t, x) = \frac{1}{\sigma_1 \sqrt{2\pi t}} \left[\int_{0}^{+\infty} q(y) e^{-\frac{(y-x)^2}{2\sigma_1^2 t}}\, dy - \frac{\sigma_1 - \sigma_2}{\sigma_1 + \sigma_2} \int_{0}^{+\infty} q(y)\, e^{-\frac{(y+x)^2}{2\sigma_1^2 t}}\, dy \right]$$

$$+\frac{2\sigma_1}{\sigma_1 + \sigma_2} \frac{1}{\sigma_2 \sqrt{2\pi t}} \int_{-\infty}^{0} q(y) e^{-\frac{(\sigma_1 y - \sigma_2 x)^2}{2\sigma_1^2 \sigma_2^2 t}}\, dy$$

for $x \geq 0$, while

$$V(t, x) = \frac{1}{\sigma_2 \sqrt{2\pi t}} \left[\int_{-\infty}^{0} q(y) e^{-\frac{(y-x)^2}{2\sigma_2^2 t}}\, dy - \frac{\sigma_2 - \sigma_1}{\sigma_2 + \sigma_1} \int_{-\infty}^{0} q(y)\, e^{-\frac{(y+x)^2}{2\sigma_2^2 t}}\, dy \right]$$

$$+\frac{2\sigma_2}{\sigma_1 + \sigma_2} \frac{1}{\sigma_1 \sqrt{2\pi t}} \int_{0}^{+\infty} q(y) e^{-\frac{(\sigma_2 y - \sigma_1 x)^2}{2\sigma_1^2 \sigma_2^2 t}}\, dy$$

for $x \leq 0$. Therefore,

$$u(t, x) = \int\limits_{-\infty}^{+\infty} q(y)\rho(t, x, y)dy,$$

where $\rho(t, x, y)$ has the form (A.21). The functions $u(t, x)$ and $u'(t, x)$ are continuous, the function $u''_{xx}(t, x)$ has jump discontinuities at the points $(t, 0)$. Applying equality (A.20) and smoothing, it is possible to get the Itô formula for the process $u(t, \zeta(t))$. Applying this formula, we get for any $0 \leq t_1 < t_2 < s$ the relations

$$u(t_2, \zeta(t_2)) = u(t_1, \zeta(t_1))$$

$$+ \int\limits_{t_1}^{t_2} \left[u'_t(t, \zeta(t)) + \frac{1}{2}\overline{\sigma}^2(\zeta(t)) u''_{xx}(t, \zeta(t))] \right] dt$$

$$+ \int\limits_{t_1}^{t_2} u'_x(t, \zeta(t)) \overline{\sigma}(\zeta(t)) dW(t) = u(t_1, \zeta(t_1))$$

$$+ \int\limits_{t_1}^{t_2} u'_x(t, \zeta(t)) \overline{\sigma}(\zeta(t)) dW(t).$$

It follows immediately from the latter relation that

$$\mathsf{E}u(t_2, \zeta(t_2)) = \mathsf{E}u(t_1, \zeta(t_1)).$$

Let $t_2 \to s$. Taking into account the boundedness and continuity of the function $u(t, x)$, together with the continuity a.s. of the process ζ, we have

$$\mathsf{E}q(\zeta(s)) = \mathsf{E}u(t_1, \zeta(t_1))$$

for any $t_1 < s$. From this equality and from the explicit form of the function $u(t, x)$ we obtain the explicit form (A.21) of the transition density of the solution ζ to Eq. (A.19).   □

It was mentioned in [33] that the normalized unstable solutions of SDEs with continuous coefficients tend to processes of the form

$$\widehat{\xi}(t) = l(\zeta(t)), \quad \text{where} \quad l(x) = \int\limits_{0}^{x} \frac{du}{\overline{c}(u)}, \tag{A.23}$$

$\bar{c}(x) = c_1$ for $x \geq 0$ and $\bar{c}(x) = c_2$ for $x < 0$, $c_i > 0$, $i = 1, 2$, ζ is the solution of Eq. (A.19). Since the function $l(x)$ is increasing, the process $\widehat{\xi}$ is a continuous homogeneous strongly Markov process with the transition density

$$\rho_{\widehat{\xi}}(t, x, y) = \rho_\zeta \left(t, l^{-1}(x), l^{-1}(y)\right) \left(l^{-1}(y)\right)', \tag{A.24}$$

where $l^{-1}(x)$ is the inverse function to the function $l(x)$, and $\rho(t, x, y)$ is the transition density of the process ζ. For the details, see [17, Chapter 3, § 9].

It is obvious that in this case $l(x) = \frac{x}{\bar{c}(x)}$, $l^{-1}(x) = x\bar{c}(x)$, and $\rho(t, x, y)$ is defined by the equality (A.21).

Therefore,

$$\rho_{\widehat{\xi}}(t, x, y) = \begin{cases} \dfrac{c_1}{\sigma_1\sqrt{2\pi t}} \left[e^{-\frac{(c_1 y - c_1 x)^2}{2\sigma_1^2 t}} - \dfrac{\sigma_1 - \sigma_2}{\sigma_1 + \sigma_2} e^{-\frac{(c_1 y + c_1 x)^2}{2\sigma_1^2 t}} \right], & x \geq 0, y > 0, \\[4mm] \dfrac{2\sigma_1}{\sigma_1 + \sigma_2} \cdot \dfrac{c_2}{\sigma_2\sqrt{2\pi t}} e^{-\frac{(\sigma_1 c_2 y - \sigma_2 c_1 x)^2}{2\sigma_1^2 \sigma_2^2 t}}, & x \geq 0, y < 0, \\[4mm] \dfrac{2\sigma_2}{\sigma_1 + \sigma_2} \cdot \dfrac{c_1}{\sigma_1\sqrt{2\pi t}} e^{-\frac{(\sigma_2 c_1 y - \sigma_1 c_2 x)^2}{2\sigma_1^2 \sigma_2^2 t}}, & x \leq 0, y > 0, \\[4mm] \dfrac{c_2}{\sigma_2\sqrt{2\pi t}} \left[e^{-\frac{(c_2 y - c_2 x)^2}{2\sigma_2^2 t}} - \dfrac{\sigma_2 - \sigma_1}{\sigma_1 + \sigma_2} e^{-\frac{(c_2 y + c_2 x)^2}{2\sigma_2^2 t}} \right], & x \leq 0, y < 0. \end{cases} \tag{A.25}$$

In particular, for $\bar{c}(x) = \bar{\sigma}(x)$ we have that

$$\rho_{\widehat{\xi}}(t, x, y) = \begin{cases} \dfrac{1}{\sqrt{2\pi t}} \left[e^{-\frac{(y-x)^2}{2t}} - \dfrac{\sigma_1 - \sigma_2}{\sigma_1 + \sigma_2} e^{-\frac{(y+x)^2}{2t}} \right], & x \geq 0, y > 0, \\[4mm] \dfrac{2\sigma_1}{\sigma_1 + \sigma_2} \cdot \dfrac{1}{\sqrt{2\pi t}} e^{-\frac{(y-x)^2}{2t}}, & x \geq 0, y < 0, \\[4mm] \dfrac{2\sigma_2}{\sigma_1 + \sigma_2} \cdot \dfrac{1}{\sqrt{2\pi t}} e^{-\frac{(y-x)^2}{2t}}, & x \leq 0, y > 0, \\[4mm] \dfrac{1}{\sqrt{2\pi t}} \left[e^{-\frac{(y-x)^2}{2t}} - \dfrac{\sigma_2 - \sigma_1}{\sigma_1 + \sigma_2} e^{-\frac{(y+x)^2}{2t}} \right], & x \leq 0, y < 0. \end{cases} \tag{A.26}$$

Let $l(\zeta(0)) = x$, $a < x < b$ and $\widehat{\tau}_x[a, b]$ be the time of the first exit of the process $\widehat{\xi}$ from the interval (a, b), i.e., $\widehat{\tau}_x[a, b] = \inf\{s : \widehat{\xi}(s) \notin [a, b]\}$. Since $\widehat{\tau}_x[a, b] = \tau_{l^{-1}(x)}\left[l^{-1}(a), l^{-1}(b)\right]$ a.s., where $\tau_x[a, b]$ is the first exit time of the process ζ from the interval (a, b), then, having respective formulas for ζ (see [17, Chapter 3, § 15, Theorem 4]), we get

$$P\{\widehat{\xi}(\widehat{\tau}_x[a, b]) = a\} = \frac{l^{-1}(b) - l^{-1}(x)}{l^{-1}(b) - l^{-1}(a)},$$

$$P\{\widehat{\xi}(\widehat{\tau}_x[a, b]) = b\} = \frac{l^{-1}(x) - l^{-1}(a)}{l^{-1}(b) - l^{-1}(a)}. \tag{A.27}$$

In particular,

$$P\left\{\widehat{\xi}\left(\widehat{\tau}_0\left[-\varepsilon, \varepsilon\right]\right) = -\varepsilon\right\} = \frac{c_1\varepsilon}{c_1\varepsilon + c_2\varepsilon} = \frac{c_1}{c_1 + c_2},$$

$$P\left\{\widehat{\xi}\left(\widehat{\tau}_0\left[-\varepsilon, \varepsilon\right]\right) = \varepsilon\right\} = \frac{c_2\varepsilon}{c_1\varepsilon + c_2\varepsilon} = \frac{c_2}{c_1 + c_2}, \qquad (A.28)$$

and for $x > \varepsilon > 0$

$$P\left\{\widehat{\xi}\left(\widehat{\tau}_x\left[x - \varepsilon, x + \varepsilon\right]\right) = x + \varepsilon\right\} = \frac{\sigma_1(x + \varepsilon) - \sigma_1 x}{\sigma_1(x + \varepsilon) - \sigma_1(x - \varepsilon)} = \frac{1}{2},$$

$$P\left\{\widehat{\xi}\left(\widehat{\tau}_x\left[x - \varepsilon, x + \varepsilon\right]\right) = x - \varepsilon\right\} = \frac{\sigma_1 x - \sigma_1(x - \varepsilon)}{\sigma_1(x + \varepsilon) - \sigma_1(x - \varepsilon)} = \frac{1}{2}.$$

Similar relations hold for $x < 0$ as well. Summarizing, in the case $c_1 \neq c_2$, we have the trespassing at the point $x = 0$ of the exit symmetry of the process $\widehat{\xi}$ out of the level ε and out of the level $-\varepsilon$, while such symmetry is specific for the Wiener process. In particular, in the case $\overline{c}(x) = \overline{\sigma}(x)$, and $\sigma_1 \neq \sigma_2$, the diffusion process $\widehat{\xi}$ was called a skew Brownian motion by Itô and McKean [24, Section 4.2, Problem 1].

Therefore, it is quite natural to introduce the following object.

Definition A.17 The process $\widehat{\xi}(t) = l(\zeta(t))$, where $\zeta(t)$ is the solution of Eq. (A.19), and $l(x) = \int_0^x \frac{du}{\overline{c}(u)}$, $\overline{c}(x) = c_1$ for $x \geq 0$ and $\overline{c}(x) = c_2$ for $x < 0$, $c_i > 0, i = 1, 2$, is called a process of skew Brownian motion type.

Let us emphasize that the explicit form (A.26) of the transition density $\rho_{\widehat{\xi}}(t, x, y)$ plays a crucial role in the introducing of the notion of the generalized diffusion process provided by M.I. Portenko in [72]. It is a Markov process whose local Kolmogorov's characteristics exist in the generalized sense.

Definition A.18 A homogeneous Markov process with transition density $\rho(t, x, y)$ is called a generalized diffusion process, if for any $\varepsilon > 0$ and any continuous finite function $\varphi(x)$ the following relations hold:

$$\lim_{t \downarrow 0} \int_{\mathbb{R}} \varphi(x) \left(\frac{1}{t} \int_{|y-x|>\varepsilon} \rho(t, x, y)\, dy \right) dx = 0,$$

$$\lim_{t \downarrow 0} \int_{\mathbb{R}} \varphi(x) \left(\frac{1}{t} \int_{|y-x|<\varepsilon} (y - x)\rho(t, x, y)\, dy \right) dx = A(\varphi),$$

$$\lim_{t \downarrow 0} \int_{\mathbb{R}} \varphi(x) \left(\frac{1}{t} \int_{|y-x|<\varepsilon} (y-x)^2 \rho(t, x, y) \, dy \right) dx = B(\varphi),$$

where $A(\varphi)$, $B(\varphi)$ are some functionals on the set of all continuous finite functions φ, wherein $A(\varphi)$ defines the generalized drift coefficient of the process, and $B(\varphi)$ defines the generalized diffusion coefficient of the process.

It is easy to see that the process $\widehat{\xi}$ with the transition density (A.25) satisfies Definition A.18 with

$$A(\varphi) = c\,\varphi(0), \quad B(\varphi) = \int_{\mathbb{R}} \varphi(x) \left(\frac{\overline{\sigma}(x)}{\overline{c}(x)} \right)^2 dx, \tag{A.29}$$

where

$$c = \frac{1}{2} \frac{\sigma_1}{c_1} \cdot \frac{\sigma_2}{c_2} \left(\frac{\sigma_1}{c_1} + \frac{\sigma_2}{c_2} \right) \frac{c_2 - c_1}{\sigma_2 + \sigma_1}. \tag{A.30}$$

Thus, taking into account Definition A.18, we conclude that the process $\widehat{\xi}$ of the skew Brownian motion type is a generalized diffusion process with a generalized drift coefficient $c\delta(\cdot)$, where $\delta(\cdot)$ is Dirac's delta function, while the constant c is defined by equality (A.30), and with a generalized diffusion coefficient $\left(\frac{\overline{\sigma}(x)}{\overline{c}(x)} \right)^2$.

In particular, the skew Brownian motion is a generalized diffusion process with a generalized drift coefficient $c\delta(\cdot)$, where $c = \frac{\sigma_2 - \sigma_1}{\sigma_2 + \sigma_1}$, and with a generalized diffusion coefficient equal 1.

It is easy to make sure that the constant c from (A.30) satisfies the inequality

$$-\frac{1}{2} b_2 (b_1 + b_2) < c < \frac{1}{2} b_1 (b_1 + b_2), \tag{A.31}$$

where $b_i = \frac{\sigma_i}{c_i}$. In particular, for the skew Brownian motion we have that $-1 < c < 1$.

We emphasize that the opposite is true. Namely, consider a homogeneous generalized diffusion process with a generalized diffusion coefficient $\overline{b}(x) = b_1^2$ for $x \geq 0$ and $\overline{b}(x) = b_2^2$ for $x < 0$, and with a generalized drift coefficient $c\delta(\cdot)$, where $\delta(\cdot)$ is Dirac's delta function, and c is the constant satisfying inequality (A.31) with certain constants $b_i > 0$. Then for the process $\widehat{\xi}$ we have the representation (A.23) with arbitrary constants $c_2 > 0$, $c_1 = kc_2$, $\sigma_1 = c_1 b_1$, $\sigma_2 = c_2 b_2$ and $k = \frac{b_1 b_2 (b_1 + b_2) - 2cb_2}{b_1 b_2 (b_1 + b_2) + 2cb_1}$.

Indeed, it is easy to see that for the process $\widehat{\xi}$ of the form (A.23) with given constants $c_1, c_2, \sigma_1, \sigma_2$ the transition density has the form (A.25). In particular, this generalized process is a process of skew Brownian motion type.

A.4 Functions Regularly Varying at Infinity

Recall the definition of class Ψ, see also Definition 4.1.

Definition A.19 Let Ψ denote the class of functions $\psi(r) > 0$, $r \geq 0$, that are nondecreasing and regularly varying at infinity of order $\alpha \geq 0$. It means that

$$\frac{\psi(rT)}{\psi(T)} \to r^\alpha,$$

as $T \to +\infty$, for all $r > 0$.

Lemma A.16 *Let the function $\psi \in \Psi$. Then for arbitrary $N > 0$ there exist constants $0 < C_N < +\infty$ and $0 < T_N < +\infty$ such that uniformly on $T \geq T_N$ the following inequality holds:*

$$\sup_{0 \leq r \leq N} \frac{\psi\left(r\sqrt{T}\right)}{\psi\left(\sqrt{T}\right)} \leq C_N. \tag{A.32}$$

Proof It is clear that

$$\sup_{0 \leq r \leq N} \frac{\psi(r\sqrt{T})}{\psi(\sqrt{T})} \leq \frac{\psi(N\sqrt{T})}{\psi(\sqrt{T})}.$$

Since for a regularly varying function ψ we have the convergence $\frac{\psi(N\sqrt{T})}{\psi(\sqrt{T})} \to N^\alpha$, as $T \to +\infty$, then for $\varepsilon = 1$ there exists a constant $T_N < +\infty$ such that for all $T \geq T_N$ the following inequality holds true:

$$\frac{\psi(N\sqrt{T})}{\psi(\sqrt{T})} \leq N^\alpha + 1.$$

Hence the statement of the lemma is proved for $C_N = N^\alpha + 1$. $\qquad\square$

According to Karamata's theorem, the function $\psi(r) \in \Psi$ for $r > 0$ admits the representation (for the proof see Appendix 1 in the book [22])

$$\psi(r) = r^\alpha c(r) \exp\left\{ \int_a^r \frac{\varepsilon(t)}{t}\, dt \right\}, \tag{A.33}$$

where $\alpha > 0$, $c(r) \to c_0 \neq 0$ for $r \to +\infty$ and $\varepsilon(t) \to 0$ for $t \to +\infty$.

Using the properties of this representation we prove the following statement.

Lemma A.17 *Let the function $\psi \in \Psi$. Then*

$$\sup_{0 < \delta \leq r \leq N} \left| \frac{\psi(rT)}{\psi(T)} - r^\alpha \right| \to 0,$$

as $T \to +\infty$, for all $0 < \delta < N < +\infty$.

Proof In fact, according to the representation (A.33), we have that

$$\frac{\psi(rT)}{\psi(T)} - r^\alpha = r^\alpha \left[\frac{c(rT)}{c(T)} \exp\left\{ \int_T^{rT} \frac{\varepsilon(t)}{t} \, dt \right\} - 1 \right]. \tag{A.34}$$

Since for an arbitrary set of numbers $r_T \in [\delta, N]$ we have the following convergence $r_T \cdot T \to +\infty$, as $T \to +\infty$, it is clear that

$$\frac{c(r_T \cdot T)}{c(T)} \to \frac{c_0}{c_0} = 1,$$

as $T \to +\infty$.

Therefore, for arbitrary $\delta_1 > 0$ there exists $T_{\delta_1} > 0$ such that for $\min(r_T \cdot T, T) > T_{\delta_1}$ the inequality $-\delta_1 < \varepsilon(t) < \delta_1$ holds. Then we obtain

$$\left| \int_T^{r_T T} \frac{\varepsilon(t)}{t} \, dt \right| \leq \left| \int_T^{r_T T} \frac{|\varepsilon(t)|}{t} \, dt \right| \leq \delta_1 |\ln(r_T T) - \ln T| = \delta_1 |\ln r_T| \leq \delta_1 C_{\delta,N},$$

and immediately,

$$\int_T^{r_T T} \frac{\varepsilon(t)}{t} \, dt \to 0,$$

as $T \to +\infty$. Taking into account the equality (A.34), we conclude that

$$\frac{\psi(r_T T)}{\psi(T)} - r_T^\alpha \to 0,$$

as $T \to +\infty$.

The statement of the lemma now follows since, as it was mentioned above, the set of numbers $r_T \in [\delta, N]$ is arbitrary. $\square$

References

1. Aarts, G., Stamatescu, I.O.: Stochastic quantization at finite chemical potential. J. High Energy Phys. **2008**(9), 018 (2008)
2. Appuhamillage, T., Bokil, V., Thomann, E., Waymire, E., Wood, B.: Occupation and local times for skew Brownian motion with applications to dispersion across an interface. Ann. Appl. Probab. **21**(1), 183–214 (2011)
3. Arnold, L.: Stochastic Differential Equations. Wiley, New York (1974)
4. Asokan, B.V., Zabaras, N.: Using stochastic analysis to capture unstable equilibrium in natural convection. J. Comput. Phys. **208**(1), 134–153 (2005)
5. Benzi, R., Parisi, G., Sutera, A., Vulpiani, A.: Stochastic resonance in climatic change. Tellus **34**(1), 10–16 (1982)
6. Bo, L., Shi, K., Wang, Y.: Variational solutions of dissipative jump-type stochastic evolution equations. J. Math. Anal. Appl. **373**(1), 111–126 (2011)
7. Borisenko, A.D.: Distribution of an additive functional of a diffusion process. Teor. Veroyatn. Mat. Stat. **28**, 5–9 (1983)
8. Carletti, M.: On the stability properties of a stochastic model for phage-bacteria interaction in open marine environment. Math. Biosci. **175**(2), 117–131 (2002)
9. Cherny, A.S., Engelbert, H.-J.: Singular Stochastic Differential Equations. Lecture Notes in Mathematics, vol. 1858. Springer, Berlin (2005)
10. Colonna, M., Chomaz, P.: Unstable infinite nuclear matter in stochastic mean field approach. Phys. Rev. C **49**(4), 1908 (1994)
11. Corns, T.R.A., Satchell, S.E.: Skew Brownian motion and pricing European options. Eur. J. Financ. **13**(6), 523–544 (2007)
12. Corrado, C.J., Su, T.: Implied volatility skews and stock index skewness and kurtosis implied by S&P 500 index option prices. J. Deriv. Summer **4**(4), 8–19 (1997)
13. Duan, J., Lu, K., Schmalfuss, B.: Smooth stable and unstable manifolds for stochastic evolutionary equations. J. Dynam. Differ. Equ. **16**(4), 949–972 (2004)
14. Dynkin, E.B.: Theory of Markov Processes. Prentice-Hall, Upper Saddle River (1961)
15. Friedman, A.: Limit behavior of solutions of stochastic differential equations. Trans. Am. Math. Soc. **170**, 359–384 (1972)
16. Garrido-Atienza, M.J., Lu, K., Schmalfuß, B.: Unstable invariant manifolds for stochastic PDEs driven by a fractional Brownian motion. J. Differ. Equ. **248**(7), 1637–1667 (2010)
17. Gikhman, I.I., Skorokhod, A.V.: Stochastic Differential Equations. Springer, Heidelberg (1972)
18. Gikhman, I.I., Skorokhod, A.V.: Stochastic differential equations and their applications (in Russian). Naukova Dumka, Kiev (1982)

© Springer Nature Switzerland AG 2020

G. Kulinich et al., *Asymptotic Analysis of Unstable Solutions of Stochastic Differential Equations*, Bocconi & Springer Series 9,
https://doi.org/10.1007/978-3-030-41291-3

19. Gikhman, I.I., Skorokhod, A.V.: Introduction to the Theory of Random Processes. Dover Books on Mathematics, Mineola (1996)
20. Gikhman, I.I., Skorokhod, A.V.: The Theory of Stochastic Processes III. Springer, Berlin (2007)
21. Harrison, J.-M., Shepp, L.-A.: On skew Brownian motion. Ann. Probab. **9**, 309–313 (1981)
22. Ibragimov, I.A., Linnik, Y.V.: Independent and Stationary Sequences of Random Variables. Wolters-Noordhoff, Groningen (1971)
23. Ikeda, N., Watanabe, S.: Stochastic Differential Equations and Diffusion Processes, 2nd edn. North-Holland/Kodansha, Amsterdam (1989)
24. Itô, K., McKean, H.: Diffusion Processes and Their Sample Paths, 2nd edn. Springer, Berlin (1974)
25. Karamata, J.: Sur un mode de croissance reguliere des fonctions. Mathematica **4**, 38–53 (1930)
26. Khasminskii, R.Z.: Stochastic Stability of Differential Equations. Springer, Berlin (2012)
27. Kloeden, P.E., Platen, E.: Numerical solution of stochastic differential equations, vol. 23. Springer Science & Business Media, New York (2013)
28. Krylov, N.V.: On Ito's stochastic integral equations. Theory Probab. Appl. **14**(2), 330–336 (1969)
29. Kulinich, G.L.: On the limit behavior of the distribution of the solution of a stochastic diffusion equation. Theory Probab. Appl. **12**(3), 497–499 (1967)
30. Kulinich, G.L.: Limit behavior of the distribution of the solution of a stochastic diffusion equation. Ukr. Math. J. **19**, 231–235 (1968)
31. Kulinich, G.L.: Asymptotic normality of the distribution of the solution of a stochastic diffusion equation. Ukr. Math. J. **20**(3), 396–400 (1968)
32. Kulinich, G.L.: Asymptotic behavior of the solution of the differential equations of the first order with a random right-hand side. Theory Probab. Math. Stat. **1**, 127–138 (1974)
33. Kulinich, G.L.: The asymptotic behavior of the unstable solution of the one-dimensional stochastic diffusion equation. Theory Probab. Math. Stat. **5**, 83–89 (1975)
34. Kulinich, G.L.: On the asymptotic behavior of the distributions of functionals of type $\int_0^t g\left(\xi\left(s\right)\right) ds$ of diffusion processes. Theory Probab. Math. Stat. **8**, 95–101 (1975)
35. Kulinich, G.L.: On the estimation of the drift parameter of a stochastic diffusion equation. Theory Probab. Appl. **20**, 384–387 (1975)
36. Kulinich, G.L.: Limit distributions for functionals of integral type of unstable diffusion processes. Theory Probab. Math. Stat. **11**, 82–86 (1976)
37. Kulinich, G.L.: Some limit theorems for a sequence of Markov chains. Theory Probab. Math. Stat. **12**, 79–92 (1976)
38. Kulinich, G.L.: On the limit behaviour of solutions of stochastic differential equations of diffusion type with random coefficients (in Russian). Limit Theor. Random Process. Akad. Nauk Ukr. SSR Inst. Mat. 137–151 (1977)
39. Kulinich, G.L.: Limit theorems for one-dimensional stochastic differential equations with nonregular dependence of the coefficients on a parameter. Theory Probab. Math. Stat. **15**, 101–115 (1978)
40. Kulinich, G.L.: On the asymptotic behavior of the solution of the one-dimensional stochastic diffusion equation. In: Stochastic Differential Systems, Proceedings of the IFIP-WG 7/1 Working Conference, Vilnius/Lithuania 1978. Lecture Notes in Control and Information Sciences, vol. 25, pp. 334–343 (1980)
41. Kulinich, G.L.: Limit behaviour of solutions of stochastic diffusion equations when the convergence of the coefficients is non-regular. In: Probability Theory and Mathematical Statistics, Proceedings of the Fourth USSR - Japan Symposium, Tbilisi/USSR 1982. Lecture Notes on Mathematics, vol. 1021, pp. 352–354 (1983)
42. Kulinich, G.L.: On necessary and sufficient conditions for convergence of solutions to one-dimensional stochastic diffusion equations with a nonregular dependence of the coefficients on a parameter. Theory Probab. Appl. **27**(4), 856–862 (1983)

43. Kulinich, G.L.: On the law of the iterated logarithm for one-dimensional diffusion processes. Theory Probab. Appl. **29**, 563–566 (1985)
44. Kulinich, G.L.: On the limiting behavior of a harmonic oscillator with random external disturbance. J. Appl. Math. Stoch. Anal. **8**(3), 265–274 (1995)
45. Kulinich, G.L., Almazov, M.: On convergence of solutions of stochastic diffusion equations with unlimited increasing drift coefficients on finite segments. In: Korolyuk, V., et al. (eds.) Skorokhod's Ideas in Probability Theory. Kyiv: Institute of Mathematics of NAS of Ukraine. Proceedings of Institute of Mathematics, National Academy of Sciences of Ukraine, Mathematics Applications, vol. 32, pp. 241–247 (2000)
46. Kulinich, G.L., Kaskun, E.P.: On the asymptotic behavior of solutions of one-dimensional Itô's stochastic differential equations with singular points. Theory Stoch. Process. **4**(1–2), 189–197 (1998)
47. Kulinich, G.L., Kharkova, M.V.: Limit theorems for stochastic differential equations without after-effect. Theory Probab. Appl. **40**(3), 469–483 (1995)
48. Kulinich, G., Kushnirenko, S.: Asymptotic behavior of functionals of the solutions to inhomogeneous Itô stochastic differential equations with nonregular dependence on parameter. Mod. Stoch. Theory Appl. **4**(3), 199–217 (2017)
49. Kulinich, G.L., Munbayeva, M.U.: Limit behavior of Cauchy problem solution of parabolic equation. Random Oper. Stoch. Equ. **2**(3), 225–234 (1994)
50. Kulinich, G.L., Petrov, I.B.: On the limit behavior of the absolute value of some of the components of a system of Itô stochastic diffusion equations. Theory Probab. Math. Stat. **28**, 79–87 (1984)
51. Kulinich, G.L., Yershov, A.V.: Convergence of a sequence of Markov chains to a diffusion type process. Theory Probab. Math. Stat. **78**, 115–131 (2009)
52. Kulinich, G.L., Kushnirenko, S.V., Mishura, Y.S.: Asymptotic behavior of the integral functionals for unstable solutions of one-dimensional Itô stochastic differential equations. Theory Probab. Math. Stat. **89**, 101–114 (2014)
53. Kulinich, G.L., Kushnirenko, S.V., Mishura, Y.S.: Asymptotic behavior of the martingale type integral functionals for unstable solutions to stochastic differential equations. Theory Probab. Math. Stat. **90**, 115–126 (2015)
54. Kulinich, G.L., Kushnirenko, S.V., Mishura, Y.S.: Limit behavior of functionals of diffusion type processes. Theor. Probab. Math. Stat. **92**, 93–107 (2016)
55. Kulinich, G.L., Kushnirenko, S.V., Mishura, Y.S.: Asymptotic behavior of homogeneous additive functionals of the solutions of Itô stochastic differential equations with nonregular dependence on parameter. Mod. Stoch. Theory Appl. **3**(2), 191–208 (2016)
56. Kulinich, G.L., Kushnirenko, S.V., Mishura, Y.S.: Weak convergence of integral functionals constructed from solutions of Itô's stochastic differential equations with non-regular dependence on a parameter. Theor. Probab. Math. Stat. **96**, 111–125 (2018)
57. Lejay, A.: On the constructions of the skew Brownian motion. Probab. Surv. **3**, 413–466 (2006)
58. Li, Y., Wang, Y., Deng, W.: Galerkin finite element approximations for stochastic space-time fractional wave equations. SIAM J. Numer. Anal. **55**(6), 3173–3202 (2017)
59. Liptser, R., Shiryayev, A.N: Theory of Martingales. Springer, Berlin (1989)
60. Luo, J.: Stability of invariant sets of Itô stochastic differential equations with Markovian switching. J. Appl. Math. Stoch. Anal. **2006**(3), 1–6 (2006)
61. Makhno, S.Y., Melnik, S.A.: Stochastic differential equation 'in a random environment. J. Math. Sci. **231**(1), 48–69 (2018)
62. Mao, X.: Exponential Stability of Stochastic Differential Equations. Marcel Dekker, New York (1994)
63. Mao, X.: Stochastic Differential Equations and Applications. Elsevier, Amsterdam (2007)
64. Menoukeu-Pamen, O., Momeya, R.H.: A maximum principle for Markov regime-switching forward-backward stochastic differential games and applications. Math. Methods Oper. Res. **85**(3), 349–388 (2017)
65. Mishura, Y., Shevchenko, G.: Theory and Statistical Applications of Stochastic Processes. ISTE Ltd/Wiley, London/Hoboken (2017)

66. Mynbaeva, G.U.: On the rate of convergence of an unstable solution of a stochastic differential equation. Ukr. Math. J. **46**(10), 1573–1577 (1994)
67. Nakao, S.: On the pathwise uniqueness of solutions of one-dimensional stochastic differential equations. Osaka J. Math. **9**(3), 513–518 (1972)
68. Nicolis, C.: Stochastic aspects of climatic transitions-response to a periodic forcing. Tellus **34**(1), 1–9 (1982)
69. Novikov, A.A.: Martingale inequalities. In: Proceedings of the School-Seminar on Random Processes, Druskininkai, 1974, part 2, pp. 89–126 (1975)
70. Oksendal, B.: Stochastic Differential Equations: An Introduction with Applications. Springer, New York (2013)
71. Pilipenko, A., Prykhodko, Y.: A limit theorem for singular stochastic differential equations. Mod. Stoch. Theory Appl. **3**(3), 223–235 (2016)
72. Portenko, N.I.: Generalized Diffusion Processes. American Mathematical Society, Providence (1990)
73. Prokhorov, Y.V.: Convergence of random processes and limit theorems in probability theory. Theory Probab. Appl. **1**(2), 157–214 (1956)
74. Protter, P.: Stochastic Integration and Differential Equations. Springer, Berlin (2005)
75. Rosenkrantz, W.A.: Limit theorems for solutions to a class of stochastic differential equations. Indiana Univ. Math. J. **24**(7), 613–625 (1975)
76. Schmid, G., Goychuk, I., Hänggi, P.: Effect of channel block on the spiking activity of excitable membranes in a stochastic Hodgkin-Huxley model. Phys. Biol. **1**(2), 61 (2004)
77. Shiga T., Watanabe, S.: Bessel diffusions as a one-parameter family of diffusion processes. Z. Wahrscheinlichkeitstheorie und verw. Geb. **27**, 37–46 (1973)
78. Shilov, G.Y.: Mathematical Analysis: A Special Course. Elsevier, Amsterdam (2016)
79. Skorokhod, A.V.: Studies in the Theory of Random Processes. Addison Wesley, Reading (1965)
80. Skorokhod, A.V.: Asymptotic Methods in the Theory of Stochastic Differential Equations. American Mathematical Society, Providence (1989)
81. Skorokhod, A.V., Slobodenyuk, N.P.: Limit Theorems for Random Walks (in Russian). Naukova Dumka, Kiev (1970)
82. Veretennikov, A.Y.: On the strong solutions of stochastic differential equations. Theory Probab. Appl. **24**(2), 354–366 (1979)
83. Wio, H.S.: Stochastic resonance in a spatially extended system. Phys. Rev. E **54**(4), R3075–R3078 (1996)
84. Zvonkin, A.K., Krylov, N.V.: On strong solution of stochastic differential equations. In: Trudy Shkoly-Seminara po Teorii Sluchainykh Protsessov, Druskininkai, November 25–30, 1974, vol. 2, pp. 9–81 (1975)

Printed in the United States
by Baker & Taylor Publisher Services